A HACKER I AM VOL 2

First Printing, 2021

Via LinkedIn – https://www.linkedin.com.au/in/craig-
ford-cybersecurity/
Via Twitter – https://twitter.com/CraigFord_Cyber
**Cover and internal illustration by Jose Herrera Velasquez
- Smartz**
ISBN E-Book:978-0-6486939-3-2
ISBN Paperback:978-0-6486939-4-9
ISBN Hardcover:978-0-6486939-5-6

FOREWORD

Writing a second book for "A Hacker, I Am" series was honestly an easy decision to make, many who have read my first book have told me they got something out of reading it. Seasoned security professionals, individuals wanting to get into the industry or people who just want to know a bit more about the topic and the industry as a whole.

That feedback gave me drive to both continue writing articles and to push ahead with a book two. Creatively named "A Hacker, I Am – Vol 2", I know I probably could have gone to some more effort in coming up with a name but I am a fan of "the Guardians of the Galaxy" series and wanted to follow their lead in the naming arena. This way I can also just keep writing more books Vol 3, Vol 4 until I run out of things to say (that's probably going to take a couple more books).

As I was writing this book the world changed quite dramatically with the COVID-19 virus that has essentially shut down our way of life in Australia and many other countries. We all found ourselves working from home and all locked up for weeks with no indication of how long this new situation is going to continue. Millions of workers now all suddenly working from home with uncontrolled devices, no security protections and IT teams so stretched thin trying to make it all happen that no one is going to notice for weeks if they do at all.

As is probably expected the malicious actors are all kicking into high gear to take advantage of the already stressful situation to steal your money, infect your machines and just be downright malicious. It's going to get interesting, to say the least, so hold onto your hats this going to get real bumpy.

Many people are losing their jobs, sadly myself included in that group being made redundant due to unforeseeable pressures on my employer with all the mandated business closures, tough times ahead for a lot of people. Makes me regret stepping down from my CSO security journalist position at the start of the month (March 2020) to look for a new platform for my voice. CSO had a bit of a restructure and it just felt like the right time to step away, hindsight may have influenced that decision or maybe I still would have done the same thing. I guess I will never know.

It is tough times ahead for us all and I feel the security threat level is only going to rise with the pandemic levels, criminals always take advantage of vulnerable situations. It's not all doom and gloom though with groups around the world like COV19 AU cybersecurity group created to help protect vulnerable health services by volunteers in this time of great burden on our health care systems. We can all come together and make a difference when it matters most, I hope to see more like this as the situation continues to unfold.

I know with us all isolated at our homes we have a lot more free time up our sleeves (I guess that is part of the reason my book is finished), so I hope that my book reaches an even larger audience and is enjoyed by everyone who has the opportunity to read it. I hope you all get something valuable out of it and learn a few things.

Keep safe and think of others during this time. Let's all come out stronger on the other side.

A Hacker I Am Vol 2

Craig Ford

Cyber Unicorns

CONTENTS

1

WHAT IF THE INTERNET NEVER EXISTED?

This first chapter is derived from a random thought I had during a presentation at the AusCERT2019 conference. The presentation was by Mikko Hypponen about security – yesterday, today and tomorrow. Firstly, Mikko is a funny guy and I loved his presentation but that isn't why I am writing this chapter. During his presentation, he said something that resonated with me and sparked this chapter idea in my head that I have not been able to shake since. He was discussing a holiday/road trip he planned in college to a lake in Europe (I don't remember the exact place) and he was discussing how at that time internet wasn't a thing at all. He couldn't remember how he would have planned such a trip without the internet, hire care, flights, accommodation and just planning where to go. All without the internet.

I know to most of us in today's society that is a completely alien idea and I honestly think most people would be lost if suddenly we were all disconnected or if the internet had never been built in the first place. Firstly, my job as a security engineer may not even exist, there probably wouldn't have been much need for it. It's a mind-boggling thought really, what would have I become if that was the case, what would our world be like, how would we communicate? Oh, wow we may actually talk to each other instead of being all cut-off and isolated in this virtual world many of us live in. I wouldn't need to write this chapter either. Let's really break this down and think about how life could have been like then maybe do a bit of reflection on how we could probably be in our day to day lives.

I am going to go down one of my hypothetical rabbit holes now that I am sure most of you are getting used to with my chapters.

Its January 1st, 1983 and ARPANET did not adopt a TCP/IP which I believe was the starting point of the modern Internet. Jump forward 36 ½ years approximately and I am working in a mechanical garage Back in my small home town in NSW. I would have never moved to QLD to pursue a career in IT/Cyber (it doesn't even exist as an idea), I love cars, so I think I would have been pretty happy in this career following in both my father and uncles footsteps. I believe without the technological advancements that were helped along by the internet, modern cars would be very similar to what they were like in the mid or early '80s. I would only hope that some of today's modern styling has come through but honestly, you really can beat some of those old classics from '60-'80s (Love the old school mustangs myself).

This hypothetical business could be a father-son setup and been running for 20 years or more now but if there are no internet customers would only be able to find us by either using the Phone book (yes they still make these – I got one a month or so ago) or they will need to actually drive the car to the garage to book in a job. No online booking forms or email requests, the internet doesn't even exist. The phone would still be something that we just use for talking on, we might have mobile phones, but they wouldn't be the smart devices we have today in which our whole lives revolve around.

Traditional mail would be the primary form of communication and is how hand-written bills for works carried out would be sent out to the garage's customers. Payments would be all made in person in probably cash or maybe checks that banks would manually record. Internet banking would not exist so there isn't

3

any other option. At least we wouldn't have those pesky email phishing scams that are plaguing pretty much every business in the world right now.

This kind of world which in some ways would be much simpler but I feel it would also make our worlds we live in much smaller as well. Without the internet, I may never have become a writer as I am today, I would never have started writing articles for CSO which helped me become the writer I am today. If we didn't have the internet maybe CSO would have been a form of the magazine we got in the mail or from the newsagents. With no internet though what would I write about? Think about it really what would you all have become if there was no internet?

Our lives would be so different. I think the internet is awesome and glad it was invented but I actually remember what it was like before it came to be, and it wasn't all that bad. I don't think I even had a computer of my own until the end of high school, yes, I had used them at school and I had a natural talent for them, but they were a very expensive item back then. They were also not a necessity so not something that warranted being purchased in my father's mind.

What about social media and all the good and bad it brings. It makes me think of the first girl I asked out, I know this will be a shocking revelation to some of you, but I actually talked to her face to face and asked her to go out with me. When I wanted to talk to her

I would need to go to a pay phone or call from the phone hanging in our hallway (a slightly embarrassing situation with your father listening in on my conversation). In the internet world we live in now they would send a message on socials like Facebook, Snapchat, Tik Tok or whatever is the latest fad. No direct contact, reduced risks, with minimal real-world social skills being developed/used.

I have concerns about what it will be like when my children are of that age when they will be using whatever platform is cool at the time, privacy will have completely eroded, and people might not even physically talk to each other anymore.

The internet has certainly changed our entire world and how we live in it. I think the developments will continue to move forward at light speed and the internet we know today will be a relic that will be talked about in museums (they will probably all be in virtual reality within 20 years as well) in the years to come. I love security and I am thank full for my experiences in it but it's a shame that something that has so much potential to make our lives even better is filled with so much hate and crime that my job is even needed in the first place. If we truly think about it, crime has always existed in some form so even without the internet criminals would still find a way to do what they do best.

The internet is a perfect place for crime, with no physical evidence like fingerprints or DNA. Just pieces of a digital puzzle that are getting harder and harder to piece back together. It's a troubling world we live in but one that is also filled with great opportunities, so I think it's time I stop talking about hypothetical what if's and get back to do what I do best and trying to make this world with the internet intact, a safer place for as many as I can.

I am not alone in this fight with so many great people out there doing what they can, we just need to come together as one unyielding front against the cybercriminals and make it so hard for them they find a new platform to try and do their dubious activities.

Sounds simple right (we should have done it by now if it was), we all wish it was as simple as that but none the less we need to try. So, with that said, let's make it happen, let's come together and meet this threat head-on. Together we have a chance but alone we will fail that is for certain.

2

SECURITY IS EVERYONE'S BUSINESS?

So, you work for a big enterprise or even a small business and you have people that look after your IT and cybersecurity needs. They have it all covered you don't need to worry about any of that stuff, that's not your responsibility or even something you understand. Cybersecurity is that weird black magic that hackers on both sides (good and evil) do in a strange and fantastical battle for your company's networks. It is like a game of strategy between medieval knights, with epic battles that are waged over the digital battlefields with blood and gore in the form of computer systems. Data is being captured, apps lost and no clear winner for anyone to see. One false move on either side could see the other take the upper hand. The battle can't be won by lone knights, this is much bigger than that it will take an army of loyal foot soldiers, squires, maidens, accountants, receptionists and sales staff.

Okay so you are pretty clever and would have caught that switcheroo I added to the end, they're right? Cybersecurity is not some fantastical black magic and there is no mythical battle for your computer systems by virtual knights per se but what I have said is 100% correct. Cybersecurity is everyone's business. Yes, I do mean everyone from the cleaners, to the accountants, to the temp workers that are only in twice a month. Everyone is responsible for cybersecurity. I don't mean that you all need to go out and get your staff configuring firewalls or advance endpoint solutions or even hunting threats on the dark web. That is crazy talk, look I know sometimes I get a little carried away and get a bit of the deep dive into that rabbit hole so to speak but I am being completely serious now.

The cyberwar is real and although it is not waged over that fantastical realm it is happening as we speak in real-time, possibly even on your systems whether you are aware of it or not. Yes, that right a malicious actor could already have you and your company in their sites. They are in the process of waging real digital war against you and you expect opponents of sometimes almost unlimited resources to be stopped by James or

Jenny your internal security person. You're not serious right? One or even five people cannot defend against all attacks on all angles at all times 24/7 365 days a year. That's just unrealistic and almost impossible. Should we just give up and not care if a breach occurs no that's not the answer, the size of the security team your company has at its disposal is the answer.

Hang on didn't I just say you only had James or Jenny to keep you secure, I did but that's not where your security team should end. How many staff does your organisation have? 20, 50, 1000 or more? Why not deputize them all into your security team, bring everyone and I mean everyone into your army. Don't try to scare them into submission, that doesn't work. Empower them. Yes, that's right to empower your teams, teach them things. Have them trust you and work with you to protect themselves. Don't have them think that Jenny or James will protect me I don't need to think about that. Help them understand the threats in plain English or whatever language or jargon they understand best. If you are an accounts firm educate them in their language they understand and will be easy for them to remember.

Look at the education process as a constant, not a one or two-time thing each year. Help your teams live and breathe security, help them to be more secure at home. Teach them best practices and also explain why they are best practices. Don't just shove new rules or more work on their laps and expect them to just do it without an explanation. You are possibly making their days harder; the process they have followed for 10 years you want them to do it differently. Staff will come together if they understand what is at risk, what we are all fighting for together. Give them a cause to stand up for and be a champion in their team.

Honestly give your team a chance, help them understand what is happening, the personal and company benefits. If you can do it right and bring on-board the right people, you know the ones, the true leaders, not necessarily the managers (they still could be) but the staff who carry influence. Get them to buy in and your teams will join in (Build it and they will come – I know it's a sad pun from a Kevin Costner movie). You get where I am going with this though right?

Instead of having one or even five security staff protecting your company, you could have 200 or 500 staff members fighting the cyberwar, being better security-conscious users, following best practise and helping raise the alarm for potential threats or incidents. I feel an army of foot soldiers and five knights could have a much better chance of surviving a battle or possibly even reign victorious if everyone comes together as one wall, one sword or one shield.

This battle cannot be one standing alone, we need to stand tall and stand together.

3

AI ATTACKS COMING TO A NETWORK NEAR YOU

Artificial intelligence is both an amazing and scary idea all wrapped up in one don't you think. True AI has been a fantasy that has been touted in many a sci-fi flick from terminator, ex Machina, I Robot, the matrix and even Wall-E (I am a bit of a sucker for kids movies) just to list a few. Some good characters and some pretty awful ones but one day (yes it's only a matter of time) AI will be removed from the world of fantasy and thrown into the mainstream world we live in. Many companies tout AI functionality in their products especially in cybersecurity but in all truth, it is only machine learning. True AI is yet to be realised but the characteristics are already starting to show themselves and when it does finally occur it will bring with it both good and bad scenarios.

It's an interesting thought if you truly think about it, there are so many possibilities of how AI could be integrated into our lives for the better (and possibly worse). Machines could solve many of the complex problems that we face and be the final piece of the puzzle needed in the search for a cure for cancer or dementia or so many other problems. Seriously machines will look at problems in a completely different light to how we humans do which could be how these problems are solved. This includes cybersecurity, we are losing the battle at this time but AI could help us react faster, respond better or even just look at the issues differently. It's a great thought isn't it, some of the world's most complex problems solved.

Now, this is about the time we throw caution to the wind and deep dive into all the AI scary movie type scenarios in which humans are deemed as less important to the newfound intelligent beings (if that is how they would be classified), humans are wiped out or enslaved Blah Blah Blah. You all know the storyline and how it goes we have seen it enough with all the Hollywood blockbusters. It's a common scenario that is thrown around and Look this is an unlikely scenario that I think is not going to come to fruition (at least I hope not anyway). I know some of you may still think this is a true possibility and you may be correct but let's ignore those apocalyptic scenarios for the moment and hope that minds like Elon Musk get the right protections in place to ensure that can never happen (if you haven't already checked out his crazy idea for human and machine integration or direct mind-machine control it's almost a mad scientist idea but if anyone could pull it off it is him).

Let's focus our attention on some real-world grounded scenarios that will soon become a real threat to our networks and society as a whole. AI-based cyber-attacks will soon become a threat that we will need to protect ourselves from and honestly it is going to be a tough job that will require us to truly think outside of the box. I recently read a whitepaper written by Darktrace on "AI-driven cyber attacks" which outlines what possible scenarios we could see in the future and what threats they have already encountered. One of the scenarios they discussed was the idea of autonomous malware (had a scary thought when I started to read that – AI-driven crypto that ex-filtrates all of your valuable data and then either encrypts or wipes out all your systems in one swipe of its AI virtual hand in just seconds – I am not looking forward to that).

Okay back to reality again, AI malware will be able to adapt and respond to your systems to be stealthier or just cause more damage. They will be difficult to detect as they will likely adapt their code as they move through systems and just be fast at what they are designed to do. In many cases, they may be finished and already cleaned their tracks before you know they were even there in the first instance. Kind of like a cybersecurity version of the boogie man. We will need to develop our versions of AI to help us detect and defend against these attacks but these tools could also be used against us by malicious actors so we will need to put strong protocols in place to ensure we remain in control of our own white hat boogie man so to speak (I am sure I could come up with a better name for it but you get the picture). It's going to be a challenge that I don't think we are prepared for.

Let's go down the rabbit hole a bit further here and consider another scenario here that was touched on in the white paper of AI-driven DDOS attacks. DDOS is a problem that any organisation that has an online presence needs to deal with and appears to be increasing in volume with a recent IoT Mirai-like botnet through almost 300,000 requests a second at an unidentified web service over 13 days, this botnet was made up of more than 400,000 devices mainly made up of home routers. Imagine if an AI entity was set loose to collect zombie devices (there will be billions of IoT insecure devices that they could utilise if we don't make them more secure), then when they reach a desired amount of devices in their army (which will be decided upon by the AI) will wage war on the determined target/s.

these targets could be critical infrastructure, hospitals, banks who knows who this tsunami will be released against and if it is true AI even the initial creator may not have that control (that's a scary thought) the AI might have just been given an objective "bring down the US or China or whoever" by all means necessary.

It will be a pretty safe attack for terrorist or nation-state actors. Set the AI beast loose and just wait for the chaos to ensue. How about I generate a scenario for you all in my usual fashion. We have a nation-state group that has set lose an AI entity that has been created to attack Australia and its daily life for citizens. It gathers an army of a billion IoT devices that are probably as powerful as a current mobile device each, it then sets its target on our power grid. Within hours we have lost power to a majority of our country, we no longer have any means of buying food or other necessities. The AI botnet then turns its attention to water and telecommunications which is already crippled due to power loss and within a day we would be descended into chaos. Within days people would start to turn on each other (Look I would hope we last more than days before people started to loot or attack each other but it probably won't), and with the AI attacking any infrastructure that comes back online as quickly as it is restored services will be hard to bring back up (it will happen but possibly not before it is already too late).

Let's say it is only down for a week imagine the wide-reaching damage that will inflict on our country, businesses will possibly fold, the economy could fall into a recession or I don't know what could be worse but it would look pretty bad. How could we stop this kind of attack though, yes we can pull the internet but as IoT is further integrated everything will rely on it and nothing will work if that plug is pulled so to speak. We will need to build in fail-safes to pull back control and prepare our systems to handle the through as pushing the pacific ocean through a garden hose. I know that sounds like a ludicrous idea but we will need to find a way, maybe with our own AI gatekeepers or something I honestly don't have a solution but with some luck, there will be a lot of people out there much smarter than me that can find the key we need to survive otherwise our idea of flying/autonomous vehicles will come to a crashing end.

I don't want to paint a negative picture here of AI and depict it as the boogie man of cyberspace but it's a scenario that truly needs our consideration before it is too late and we have lost control of our networks. Let's start the conversation and put some brilliant minds on this task, I know it will help me sleep better at night. So as usual tell me what you think, laugh at me if you like and tell me I am losing it if that's what you think but let's do something to help fight this future threat before it's too late.

4

ATTACK PLAYBOOK FOR TARGETED FINANCIAL SCAMS HURTING SMALL BUSINESSES, SO YOU KNOW WHAT TO LOOK OUT FOR?

I don't know about the rest of you but it's hard not to get worn down on the constant bombardment of news about cyber breach after cyber breach. Every day more victims are coming forward or being exposed on public forums. Personally, it's always better to admit the issue yourself than be outed to the public by a security researcher or hacker but that isn't always an option especially if you don't even know that it's occurred in the first instance. I don't want to paint a completely bleak view here though as people are doing some great things that are helping to reduce the issues that we are seeing in the security trenches daily which is good news.

The human side of security is coping a big hammering at the moment and is getting a bit of a bad name which I don't think is entirely fair to those users who are being tarnished with that negative brush. I think it is on us the security folk who are here trying to protect them and their systems. We have a bit of a reputation that depicts us as the "no" people, the party poopers who lock down our systems and don't let us have all of the new technologies that would help them do a better job. We need to change that opinion and be the "yes" people. Don't just say no all the time but think about what they are asking if what they want isn't going to work because of security concerns find an alternative that will give them what they require or at least close to it.

I know I am a little off-topic and I haven't even talked about the chapter topic yet but I just wanted to put it out there that we need to become more approachable in our respective organisations, to the community or anyone else we help to keep protected. That way when people have concerns or see something that they think isn't right they will come to us about it for help (you see why I have given that intro now).

This type of openness will make our lives as the protectors much easier because we won't be those painful "no" people anymore and it will reduce the shadow IT problem as well. If we help them get what they need there will be no need to try and circumvent us to get what they need to do their jobs well. It seems like a no brainer right, we help them get what they want, and then they involve us in getting new systems that help us do our job better.

Now once we have built this relationship with our users/people we protect we need to do more by helping them understand how attacks come through and how they can better protect themselves and the organisation. Teach them user awareness training and map out threats for them so they understand. Not just be aware of the threat but understand how an attack will take place. Lay it out for them so they know what to expect.

Let me run through a scenario to show you what I mean. How about one of those financial scams that we are all seeing getting around via email:

1. A malicious actor finds a target company and employee usually the accounts person. They could have obtained the person's info via a

previously breached company that this person deals with (happens with a lot of office 365 and Google mail accounts), they get access to an account and then go after all of your contacts. It's a successful technique.

2. They would likely send an email from either a breached account or a newly created account with say the target companies CEO as the email address or CFO like CFOName@gmail.com or something like that. The email would sometimes start with a general statement that they are out of the office and don't have access to work email and would like to know how long it would take to do a money transfer to such and such country.

3. Other times the malicious actor may redirect the email from a victim's inbox and then modify payment details on an invoice, then drop the modified invoice in the inbox again. Many organisations don't validate these requests and will just update the account info before paying the invoices. They may even add a note that the bank account has been changed just to make it look more valid.

4. A few weeks later the company will follow up with you about payment and you will indicate that you have already made payment. This will go back and forth for a few days maybe until one

of you realise that you have paid the invoice into the wrong account and they had never actually changed the bank account. You have just realised that you have been scammed.

In this type of scenario, we could have done a lot of things to prevent it from taking place. I will list a couple now:

- Multifactor authentication would certainly have helped protect the email account from the malicious actor gaining access but that wouldn't help if they were using a generic impersonation email address but good to think about anyway.
- Policies, procedures.... Make sure you have them and follow specific verification procedures before any account details are changed for payments. Call known contacts and verify that the request is valid, have a second person validate the request with you if possible.
- Training, training and discussions. What I mean is let's teach our users what to look out for not just once off training but regularly, have them send suspicious items through to you to look at. A few minutes of checking an email is nothing to help protect your systems. Give praise to people who do this, make it a positive thing in your organisation, help them learn and know

they can come to your team for help anytime that they aren't an inconvenience. They might be the key to saving you at that critical moment that could prevent a breach or financial scam like this from occurring.

Look I know what I am saying above is nothing new and I am covering over stuff that many of you probably already know but we are very bad at the human side of security. We need to do better and work together more. That's the only way we can win this battle against an unrelenting attack.

Pull up your socks and get ready to prepare to do the hard yards to help your humans do better at protecting the company and themselves in their daily lives it will be a benefit to everyone. Oh and don't forget your technical people, they may be better at seeing these scams but we can all fall victim to a scam in the right scenario. So let's educate everyone and don't be judgemental it achieves nothing for no one.

5

CHILDREN'S SMARTWATCHES, A REALLY STUPID IDEA

As a parent, I feel that we need to think about our children's safety, don't make rash decisions based on un-validated information that will leave us looking red-faced or our children in unsafe situations. Many so-called smart devices are being created for use with our children that are a form of a digital nanny. Smart teddies, baby monitors and smartwatches just to mention a few but these devices are bad for both security and privacy. I have discussed some of these items in my previous chapters, but I want everyone to understand how bad some of these devices can be.

I have seen examples of devices being hacked and malicious actors holding the owners of them to a form of blackmail, watching and listening in on devices without the owners knowing anything about what was occurring. These devices are just absolutely crazy, they are cheap low-quality devices in most cases with minimal or sometimes no attempt for security. I just don't understand why people use them, they put you and your families at risk. We need to really think about these smart devices before just rushing out and buying whatever the latest fad device is, do some research know what you are giving up (I am not talking about money here), know the risk, know what information the device collects from you?

Did you know that Amazon (I am sure the rest do it as well, they just haven't been caught out yet) has people that listen to recordings that their smart assistant makes for so-called research to improve how the assistant functions? I am not sure I completely believe that but let's just pretend I do. Look I am not ignorant and I am sure that none of you is either, we all know that these devices record us and listen to everything we say that's their job, it's so they can respond when requested but I think that Amazon, Apple, Microsoft, Facebook, Google and all the other monstrosities that are tech companies these days need to be honest and forthright with what they are collecting.

It shouldn't be something that is dragged out of their dirty laundry for all the world to see when they are caught but something that they honestly say to their customers "Look we listen to some of your recorded conversations on our devices so that we can improve how Alexis works, are you okay with that?" hey maybe even have an opt-out option if you are not comfortable with it and actually stick to it. Just my thoughts on this but I am getting a little off point here.

Let's look at a whole other level of dangerous IoT devices, **Children's smartwatches**. These devices just scream danger to me, this may be the hacker side of me clawing to the surface, wanting to do my worst and bring these devices to their knees but I just can't do it. These smart devices are strapped to children's arms, a personal band that says, "kidnap me, I am right here". I know you are all probably thinking wow Craig you are being a bit dramatic today. Yes, I might be, but I want you all to really think about the risks here not just brush it off in the Australian way "Shel be right" this is our children we are talking about here so that just won't cut it this time.

These devices use both mobile and GPS tracking services to monitor and communicate with your children. Any time you want to know what they are doing or check the tracking to know where they are just open your app and there it is. If you can do this though don't you think a malicious actor could as well? These are cheap devices with no real interest in security, so what is there to stop them. Many parents who will purchase these devices will think they are making their children more secure but in fact, you could be doing the opposite.

Some smartwatches have what they call a geofencing option in which you can allocate a virtual boundary with GPS markers that is the defined safe zone for your child and if they venture out of that allocated area it can alert you to this infraction but many of these devices will not alert as they are designed to do when the device leaves the geofence. So, if you want this device to let you know when your child ventures too far from home it may let you know 8/10 times as it is designed to but what about those two occasions? (or more this is just a hypothetical). Honestly, if this was my child I would not be happy with an 80% success rate (100% per cent or nothing in my opinion).

What about the GPS tracking function itself, I saw Troy Hunts talk at the AusCERT2019 conference (I mentioned this in my chapter about the conference), during his talk he told a story of when he allowed one of his colleagues to manipulate the GPS tracking of his daughter's smartwatch (I don't remember the specific device but that doesn't really matter) via a flaw in the API call, the devices and its control application used unique identification numbers for the smart device. The problem was that you could modify the identification number you wanted to connect with and once you were authenticated it didn't do any further checks (I think that is an Ooops moment on the manufacturers/designer's behalf).

So, they then changed the code in the control app with an instant connection and control of Troy's daughter's smartwatch. This allowed them to move the GPS location from her playing tennis somewhere on the Gold coast where she was meant to be, to suddenly be out in the middle of the ocean. If that was your child and you didn't have a friend who was telling you what they were doing this situation would be terrifying for a parent.

The same flaw that Troy discussed also allowed for a malicious actor (not so much in Troy's case) to also call the smartwatch and communicate directly with his daughter. This functionality is supposed to be locked down to just the parent's account, but this particular app and device have some serious security flaws which still might not have been fixed (I truly hope it has though).

I would like you to consider this a moment, a malicious actor knows where your child is, they can listen/communicate with them anytime they would like to. If they call your 7-8-year-old on this device, they will not question a malicious actor if they say mummy or daddy asked me to come to collect you for whatever reason they decide. They don't have to say where they are (they already know that) and when they pick them up they can make you think they are on the other side of town or at home or anywhere they want you to think. This sounds like a scary situation to me and I just want people to really think about the dangers.

I hope that I have made you all consider the risks involved with these fancy kids GPS trackers and you will all take a step back to figure out if they are something that you are comfortable with for your children. I know I am not, especially the really cheap ones that are not even considering the security element, they just focus on the core functions (which don't always work well) and keeping the costs down so more people will buy them. It doesn't matter to them if the purchasers are oblivious to the risks, they just want to make a profit on their devices which I understand but it should be at the cost of safety to the children who wear their devices.

Look I could talk about the risks on these devices for another 1000 words but you know what I am trying to tell you, do your research, don't just leap and buy the next new toy. Consider the risks and go into them if you decide to with open eyes. Don't have blind faith in the devices, actually be parents to your children and don't leave their safety to these insecure and unreliable smart devices, trust me if you do you could really regret it.

That's enough from me for this chapter I feel but as always tell me what you think about these devices? Do you think I am being over dramatic, or should there be some sort of regulations introduced to ensure safety and quality standards are upheld? I want to know your opinion's, as we need to make this world of ours more secure together, it's not a battle we can fight alone.

6

COULD YOU BE
HACKED BY YOUR
PRINTER?

You have a very secure environment, you have the latest firewalls which have been configured well, you have really good network segregation, and you have the latest in endpoint detection and response platforms with a well-configured siem platform. You even have a patch management system that is making sure that your systems are all up to date and not vulnerable to any known threats or vulnerabilities. Wow, you are really on top of this and are doing great at ensuring your systems are well protected and managed. Nice work, you should be quite happy with your progress but can I ask you a question? Did you put up 10-foot walls and secure your entire environment only to leave the back door unlocked and ajar for all to enter by port 9100?

Network printers are a huge part of most organisations and homes in many cases with direct internet-connected printers and Wi-Fi hotspot configurations to allow for simple\easy and reliable access to modern IoT printers. Yes, many printers are now connected to the internet especially home units as they are made to be accessible to users via mobile apps to help manage printing and enable direct printing capability for most non-technical users.

Printers in organisations wouldn't be connected to the internet though, would they? That doesn't sound correct? – That is what you are thinking right? This is only an issue for home setups, not my organisation? I just did a search for internet-connected printers in Shodan and the results were 30,532 with 318 available in Australia (There is also a whopping 8,910 available in the US). On the first page, I found two printers sitting on Australian university networks ripe for the picking. Seriously I am not even putting in any major effort to find this information and it is freely available to everyone. Admittedly universities are known to make printers available like this and in my opinion is a massive security risk that should be resolved.

So what you say, why do I care? What could someone do to a printer on my network except drop a million print jobs or put a print loop with a specific message like has been done recently from malicious actors saying that you have been hacked and to pay up via bitcoin? Yes, that is certainly one attack vector and I have seen it work very well surprisingly. That's not a concern for me though, just a pain in your side maybe, a bit of an inconvenience. My concern is this.

Let's say a malicious actor or even an internal actor wants to get all of the print jobs sent through to the finance printer or even HR? Could they manipulate a printer to collect copies of all documents that are sent to the printer? Yes, that is quite a simple attack that could be carried out against network printers if you are on the same network as the unit. This could be done for an external attacker by breaking into your organisation's Wi-Fi and then completing an attack on the printer using a tool called PRET. The process is quite simple and very effective. You can imagine the level of data that could be captured by this process. Do you want to know what salary or bonuses that other staff get? Do you want to get personal information, this could be your perfect method to get it.

Denial of services and print job-stealing, in my opinion, is not of major risk to the organisation but is something that should be considered. It is possible to mitigate these attacks with traffic flow control that only allows access to printers via certain IP's or users but it will need to be determined if the risk is worth the elevated configuration requirements.

I think the real risk is the printer being used to attack systems on your network. Yes, that's right, it is possible to use network printers to attack other machines on your network and allow a malicious actor to take control of a workstation or server. I am not going to go through the technique as that is not what the chapter is about, I want to demonstrate to you all that network printers are a security blind spot and we don't manage them well. We need to manage these as we do other network devices. Secure them first by not leaving default credentials (please at least do this – most don't change them), update the firmware/software for the printers – these are used to patch vulnerabilities and is a great way to help reduce the attack surface. Don't connect your printer to the internet if it is at all avoidable (please), this is an unnecessary risk that you don't need. Don't allow the printer to be used for unauthenticated SMTP email traffic, this could be how you get your latest malicious phishing email from someone pretending to be the finance manager. Only allow authenticated email communications and only allow them to be sent from authenticated users.

I know that in some instances it's necessary to allow printers access to the internet but restrict the access and make sure you understand the risk, monitor access and be smart about how you allow it to be used.

Otherwise, you may be on the morning news being the latest breach victim all because of that blind spot you call a network printer. Seriously network printers need to be considered as a risk and appropriate security controls put in place. Don't do the standard Aussie thing and say "it'll be right", plug the security holes, patch devices and let's all have a great start to the New Year?

7

COULD YOU THINK
LIKE A HACKER?

It's 3 am on a Wednesday, I am scratching my head, starting to get a bit stressed. I need to find a way in. I get up and make myself another cup of coffee, this must be like my 10th cup tonight or morning now. I grab a teaspoon from the draw and spoon in a spoon of coffee and two sugars. I stand and wait for the jug to boil and decide that a snack would go down well so I opt for a packet of crisps the big family size bag. Probably not the best choice for my health but hey this is already my 10th cup of Joe, I don't think the chips is the issue.

I pore in the water and give it a quick stir before heading back to my desk. I look over my log and I realised I have been at this for 12 hours, I am not going to let this thing beat me. There must be a way in, there is always a way in. I have a mouthful of coffee and then eat a handful of the potato chips, they are extra salty just how I like them. I ponder for a moment on my task, my opponent. I know them, I have researched them. I know everything I need to know to get through the security measures. These people don't even have two-factor authentication, this should be a walk in the park.

I look over at my other screen and see the social media page belonging to one of the targets. I see a cat with a name collar around its neck that says Bubbles. I look a bit further and they have a daughter who is at turning 18. Wonder if the password will be the cat's name and daughter's year of birth, it couldn't be that easy after I have spent 12 hours trying to break in. I punch the credentials in, fail. Grrrr, this is getting frustrating. Mmmm wonder if any of the users are using the same passwords for any of the account breach passwords. Some of them are on LinkedIn and they had a breach dump file with all of those passwords. Surely, they would have different passwords or at least have reset it by now. I search my dump file and pull out three.

I enter the credentials on my script to be piped into the server login screen and the first one fails. The second one fails. Boom I got you. The third one was successful. I am in and I have access to almost 80% of the files. Maybe the 12 hours was worth it and today will be a good payday.

Okay, okay that's enough of the fantasy hacker stuff (although I like writing it), I am trying to set the scene for a thought or an idea that I want everyone to do. I want you all to think a little more like a hacker or malicious actor (I don't like to depict hackers as bad people as I am one myself in a sense and I think I am a pretty good guy). So, what I want you all to do is think before you do something. If you want to plug in an access point into a network port so you can surf the internet on your iPad in the office, stop and think about it for a moment. If you plug in an unsecured Wi-Fi access point into the company's network is that the smart thing to do? Could a malicious actor use that to easily gain access to the system, yes, they certainly could?

What about passwords, if you were looking over your social media could you put together what your password may be from your photos or posts? Don't use family member's names, date of births or pets. It's a bad idea. Pick some random words, put them together to make something that sounds humorous to you and add a couple of numbers and a symbol. Pretty hard to crack the password, pretty easy for you to remember.

Don't plug a random USB drive into your pc or use a stranger's charge cable. All of these things could be used by you the imaginary malicious actor (hacker) to break into your systems so why would you use them? Why would you install random software on your computer that hasn't been vetted for security threats? Yes, I know it takes time, you don't have to get IT or security to check these things for you but you must think more like a hacker, be more suspicious. Think of everything as a possible attacker or attack method and your malicious thinking just might save your company from a public embarrassment from a breach. Big breaches can destroy great companies so put on your hacker cap and consider if what you are about to do is the right thing. Trust me I am a hacker; would I steer you wrong.

Okay in all seriousness, this is a good idea. Do it, it will help you understand threats and better protect yourself.

8

CYBER SECURITY IS IT SPELT LIKE THIS OR IS IT CYBERSECURITY? WHY CAN'T WE DECIDE?

The cyber security industry has been around since back when the internet as we know it was brought to life in 1983 when they switched the TCP/IP on for ARPANET (probably even before then). Oh sorry, it would have been called "Information Security" back then and guess what it is still the same thing. Some people somewhere wanted a more marketable name, so they started to call it cybersecurity or cyber security.

The sad thing is that this new buzz name for our industry that has been around for a couple of years or more now, we are all arguing about how we bloody spell it. Honestly, I don't really think it matters that much and we should focus our attention on more important things like how we can catch cybercriminals in our network faster (they are probably already in there we just have to catch them).

Never the less, I keep seeing this topic on social media and it was at the AusCERT2019 conference last month. They actually had a vote on this in one of the talks (I don't remember which one specifically) where they asked people to raise their hands for which way they thought it should be spelt. The "Cyber Security" two-word method for spelling it was undoubtedly the winner in that vote with I would say more than 70% in its favour. As many of you would have seen in my chapter I was voting on the opposite side of that vote as I have always written it "Cybersecurity" in all my writing but if the industry thinks it is two words, not one then who am I to say otherwise.

From this moment forward I will try to remember to use two words not one in my chapters or any other writing that I do (I may slip up now and then. I am not perfect). Okay, so that is my pledge lets quit arguing about how to spell this buzz word name that has been given to our industry and start to focus on the more important things like what brand is best for security professionals to drink to ensure maximum efficiency. No, what about the best security platform to use to protect your systems (that's certainly more important) but even that doesn't really matter.

Each person will have their own opinion on things, the best platform to use, favourite colour, favourite brand of shoes, best all-time car brand or car (Old school mustang fan myself). Does it really matter if I like Apple phones and you like Android it means nothing, seriously nothing at all? BY now I am sure what I am trying to point out to you, but I will cover it a bit more just to ensure my point gets across to everyone.

In "Cyber Security" or "Cybersecurity" whichever way you like to spell, we have so many problems that we need to focus our attention on resolving as a community like diversity, breach after breach with no sign of slowing down with no sign of a solid solution to stop it (I still think we need to all do the basics better that will certainly help). Maybe we could work out how we can all come together as a united front to fight the avalanche of attacks, together we may stand a chance.

So, what I am saying is let's forget about how we spell a silly marketing name for our industry and either just straight out pick one (which will probably never happen) then get on with what is important. Protecting all the systems we are collectively responsible for. It's a big job and will require our full attention and is obviously more important than how we should spell a word don't you think.

What about diversity? Diversity is a massive problem in our industry with a very low percentage of female members coming into and staying in the industry. CSO, as well as many other organisations, are already focusing on this problem with voting recently closing for the "women in security" awards which is a great initiative, there are so many initiatives I can't mention them all but shouldn't we focus more attention on this? I think we should, maybe we could even use some of the excess energy we have to find a way to go beyond diversity for men and women but race, background, education and more. We want to have as many different viewpoints as we can to have any chance to be successful in really making a difference in our industry. Let's not get too distracted and aim for true diversity in all forms, it will be a benefit to us all.

What about collaboration? I mean real collaboration in our industry to help share intel, help each other learn how to better protect ourselves? I have covered this before in previous chapters and I will cover it again because it is important. We need to look beyond competition and try to be better together, if we can see past the possible roadblocks that keep us from working together we could make a difference. I am part of collaborations that bring together people from very strong competitors but, in that room, we are not enemy's anymore, we are there for the same goal "protect our clients and our networks" a truth that forgets everything else.

So instead of wasting all our time arguing about a name let's put this to rest now and put our attention into one of the more important problems that we need to focus on. What do you think? Can we do that?

Now that I have got that rant off my chest let's get back to cybersecurity like we should be, oh bugger I have already messed up and spelt it wrong. I will try harder ☺ maybe. As always let me know what you think, do you care about how this is spelt and think it is worth our time to sort this out? Should CSO run a poll on the correct way to spell this and announce the winning version? Seriously tell me what you think, I want to know but let's not waste any more time on this than is needed. If we want a decision let's take a vote on this and put it to rest once and for all.

9

CYBER TERRORISM IS A REAL THREAT

Modern terrorism and cyberwar are major threats to our society's way of life. Gone are the days in which acts of war or terrorism need to be carried out by foot soldiers on the ground. Serious pain can be inflicted on targets without ever stepping foot inside the country in which they reside. Attacks against governments, businesses or critical infrastructure can take place in seconds with well-planned synchronised attacks. These attacks could bring down power, telecommunications, financial or even transport. Seriously imagine the chaos if all trains were stopped, traffic lights were all set to green or red or even just flashing yellow in almost any major city around the world. That would be chaotic with a massive financial impact on the economy.

Let's think about it, Sydney Australia has approximately 5.23 million people, with approximately 1 in 5 people (the statistics are a few years old but will demonstrate my point) who use a form of public transport to get to work each day. That's 1,046,000 people making their way to work that suddenly won't turn up. Just imagine the complete standstill that will cause thousands of businesses to close their doors and many more to try and run at a greatly reduced capacity. Can you imagine it If $1/5^{th}$ of all your staff didn't turn up today? I think the 1 in 5 not arriving is probably being generous, if public transport was out, many of those 1 million people would then try to get to their jobs via road which will cause massive further delays in commute for a city that already suffers from severe traffic jam issues.

If malicious actors then crippled the traffic signals as well, I think it would be easier for everyone to just close their doors for the day. What if the problem couldn't be resolved in a day, people who had made the extra effort to get to work may not be so keen to continue the process if it increased their days by an extra even 2-3 hours. A 12 hour day instead of an 8 hour one and the workload would be hideous with a large percentage of absent staff. Longer hours, reduced staffing, massive work that sounds like a recipe for half of Sydney's population suddenly getting a case of the flu – Cough, cough.

Seriously, we are not ready for a full-scale cyber-terrorist attack like these, I don't think it is even something that has crossed most people minds but it is going to happen that is for certain. We need to consider the impact and have plans in place that we can enact to ensure that our businesses can continue to work as needed.

In some cases, we could look at having the ability to ramp up remote access capabilities to allow stranded staff to work from home or another non-city central location required? This is smart planning and would be a very useful addition to include as part of a disaster recovery plan. Look I get it, you are all probably thinking why would I plan for something that will probably never happen or even if it did wouldn't affect you or your organisation.

You need to be prepared for the possibility and at least know what you would do if it does occur. Yes, you are more likely to have a fire or be affected by a severe storm but cyber terrorism is a real threat, along with cyberwar, so you need to ensure that as part of that disaster recovery plan you have a specific process that can be enacted in an instance of cyber terrorism or cyberwar. If you don't I believe that you are leaving your organisation open to a threat that could be planned for.

Somethings are hard to plan for but it doesn't mean you shouldn't at least try to manage the risks. So please go include this threat/recovery planning and make sure that mock scenarios are considered thoroughly it's the only way that you can have some ounce of preparedness to reduce the possible impact.

So as always, tell me your thoughts, share your expertise and help us all be better prepared for any threat we might face in this truly unpredictable landscape we operate in.

10

DEEP FAKE IS THE GOLDEN EGG FOR SOCIAL ENGINEERING ATTACKS

Social engineering attacks are probably one of the most used attack vectors that customers face in today's chaotic and often very challenging technology landscape. I have talked about this in previous chapters and my book but I sadly have to tell you that there is another method that is being utilised by malicious actors that are going to be hard for people to pick up.

Deep fake technology is a bit of a hype technology that has been getting around for around a year now and was in my opinion just a bit of fun with no real threat to businesses with regards to information security as it didn't really work properly and was easy to determine that it was a fake voice or video that was generated by a computer system. However, that is no longer the case and several instances have come forward in which malicious actors have used the voice creation part of deep fake technology to generate voicemail messages or have a generated conversation on the phone with employees of a company indicating that they were the CEO or finance manager. They would request payments to be made or account details to be changed.

In all of these instances that I am aware of these deep fake voices were used in combination with an email sent in a similar method to many we have all seen before from an email address created on generic hosting platforms with the CEO's name or whoever they were impersonating. This would give the email validity if they heard what appeared to be the voice of that person on voice mail or a quick sharp call that ended quickly (so the person couldn't ask questions).

There are many versions of the technology available with one of them being Lyrebird which does a great job of creating what they call a voiceprint or voice map. What happens is very similar to the process that you go through when setting up your google home or Siri or whichever personal assistant has its tentacles wrapped around your particular platform. You are asked to say some pre-mapped out items in which it uses to generate your voice patterns so that it can easily mimic you.

I know what you are thinking how can they do this with someone that is not voluntarily saying the required phrases? I'm glad you asked. It's easy actually in many cases as CEO's are not normally shy individuals and will in many cases have video recordings of them on the internet from interviews or company promotional videos or whatever source they can find. If that isn't the case they could easily infect a device with webcams being quite an easy part of a system to gain access to on many occasions (have a google for recent large scale breach allowing access to thousands upon thousands of webcams) and then just record/process the voice recording to get what they need.

Honestly, this method is harder because you need a much bigger sample of the person's voice, something in the vicinity of 8 hours but if you have access to the person's webcam I am sure you will be able to get it. We are all at risk of this threat, not just our users. Would you say no to your CEO if they requested you do something now for them? Most wouldn't honestly ask some people in your office what would they do?

Let's layout a scenario, the CEO calls them and leaves a voice mail says they need a payment of $$$$ made to this account in this person's name today for a new project ****. Many of them if they heard the voice and it was the CEO or financial controller they would ignore any normal checks and just do it. Seriously ask them. I know we have all been working hard on getting people to have a secondary check process, maybe even a third but this blows the normal logic out of the water. We would normally instruct users to call and verify that this is correct before doing changes like this or making irregular payments but they just heard the CEO himself tell them to do it. That wasn't them in this scenario but how would they know that? They wouldn't this isn't something we have warned them about before but we need to ensure that they are aware of this possibility and we need to adapt our procedures to ensure that this threat can be managed.

Still not convinced that this is a real threat? Check out this video from Journalist Ashlee Vance. It was almost perfect with replicating his voice and it was from over ago. Technology has come a long way since then and in my opinion, it was pretty good already.

So now that I have you all convinced it is a real threat, what can we do to help stop this threat? I think we just need to change up our procedures a little to cope with this new attack method. If a user receives a voice mail and email from someone appearing to be the CEO or Finance manager or whoever, staff need to take the information at face value only. Once the call has finished, email the internal company email for the correct person, call them on already known numbers to confirm that the request is real. If they are not available then take the request to whoever is in charge in their absence to validate if it should take place or not and they can make the final call.

If you can't verify the request then your procedures should allow the staff to decline the request no matter who the person is doing the request. All requests should follow correct company procedures with no exceptions to that rule. I know people don't want to get in trouble with the CEO but in cases like this senior management needs to back up the procedures and there should never be consequences for staff following correct procedures that is their job after all.

It's a scary new world we live in and it's getting harder to know what you can and can't trust but I think if we create solid procedures and stick to them we will be okay. Don't just push this to the side though and think this won't happen as it will and you need to be prepared. A little bit of preparation and you won't have egg on your face when it does? You can thank me later.

11

DIGITAL IDENTITY REPLACING PHYSICAL ID CARDS

I remember growing up in the '80s and 90's seeing lots of movies where teenagers would try to make or buy fake id's. Particularly in the US where the legal drinking age is 21. ID cards such as a driver's licence, which was originally introduced back in 1910, have long been physical cards that we would carry around with us. This was to prove who we were and our age for the younger members who wanted to go to pubs and clubs. They started as little more than a piece of paper; I remember my fathers while growing up had a laminated licence with a photo and some details.

When I finally got my learners in the tail end of the 90's we had moved on to one of the first versions of the hard plastic licence that we nearly all have today. Change is afoot however with digital ID's starting to become a reality, the Queensland Department of transport and main roads launching the first piolet of the digital licence app late this year (2019). They are calling it a digital wallet and it is essentially a digital identity app on your mobile phone that will be able to be used as a replacement for the old-style physical card.

They have added a security feature that is designed to be able to keep your information secure by utilising the security features of your phone with a pin, fingerprint and possibly facial recognition functionality. On the TMR site, they indicate that the new digital licence will be more secure than the traditional form of ID and you know what I think they are probably right. There is no real way of securing the traditional licence or physical ID card.

It can be used as your driver's licence and as proof of ID. During the piolet stage, they are not making it compulsory for cardholders to move over to the new form of ID, if you choose, the traditional ID cards will still be made available. QLD is not the only state rolling out with these test digital ID's with I believe most states in a similar phase.

It's not just your licence going digital either with DFAT (Department of Foreign Affairs and Trade) working towards a digital passport however they are not quite at the stage of trialling these just yet. Australia Post has also jumped on the digital ID bandwagon with the release of the Digital ID service that will allow you to pick up packages and also use this service at participating other organisations instead of needing the old school licence anymore.

Look, all of these services have some good features and the idea of a digital ID is a great idea but is the technology at a standard that can provide a safe and secure service. What happens when a malicious actor finds a vulnerability in these platforms (oh and let's be very clear here they will), what information will they be able to gain and what would they be able to do with it?

As I started to write that last line my mind exploded with possibilities (that's my hacker side coming out there), I do not doubt that they will be able to steal all of the licence holder details, manipulate the apps to create the modern version of a fake ID. The fake ID scenario will be an interesting one and it will be interesting how they ensure the validity of the ID's as I am sure with the technical skills of some of the younger generations at least a few of them will have the skills and know-how to be able to throw together a decent cloned version of the app.

If there is no way to validate this on the fly at night clubs or similar locations and to the people checking the ID's the app looks the same and displays/

does all the right things it will won't be long before the backroom trade of fake ID's will start to form. It's an interesting thought but I would assume that these services will have some good validation processes and will have easily available methods to validate the ID being provided. It does come back to the likely vulnerability though, will malicious actors (or one of those resourceful teenagers) be able to change details on the displayed ID and will that then still allow them to validate against the verification methods. I think there will be a way to do it and in my opinion, it will only be a matter of time before this will take place.

I think this discussion is important for us to have early as these services will inevitably flow over into the business world and we will all be the ones who will be tasked with securing them for our organisations. So how about we get ahead of the game and start a conversation on this and let's ensure that as an industry we are prepared for what is to come with Digital ID's and whatever is to follow that.

12

EDUCATING CHILDREN ABOUT ONLINE RISKS: DON'T BELIEVE EVERYTHING YOU SEE

It's scary how little parents know about what their children do online. Do you know exactly what your kids do? I see children as young as 3-4 years old given tablets with no parental controls, no restrictions, and no protection. They can download and install whatever they want, they can look at anything they want with no restriction.

This is the first problem that we need to resolve. Do not give your young children electronic devices that they have uncontrolled access to without having parental control apps on them. Keep control of these devices and know what is happening on them. That way you can understand their usage levels, what they do and what platforms they communicate on.

Oh and please don't let young children create social media accounts, this is crazy. Why? I know some of you will want to know why this is a bad idea so here goes, let us run through a bit of a scenario so it makes sense.

Let's say we have a 10-year-old girl - Sophie. Sophie is a great kid, she does well at school, never gets in trouble and is very well mannered. She has an iPad and has no restrictions on her access to the internet and can download any apps she likes. Why would her parents restrict her? She's a great kid, they don't think she will do anything inappropriate. She has set up an account for Facebook to talk to her friends and share pictures and things. Seems pretty safe so far but what happens if she is approached by someone on social media that is pretending to be someone her age, over time convincing her to send pictures to them or even worse getting them to secretly meet them somewhere? I hope the alarm bells are starting to ring now. This is a real possibility.

This type of scenario happens too often and we all need to do our part to make sure our children are safe online and pay attention to what our children do online. Yes, I know this is a worst-case scenario type incident but they are more common than you think. Set up parental control features restrict times of use and pay more attention. More needs to be done.

You need to sit down and talk to your kids about the risks of social media and the online world in general. I know these types of talks are difficult and we don't want to scare our children but it is important they understand that there are real risks. They need to know that online things shouldn't be believed without further validation. Just because someone says they are a 10-year-old girl or boy doesn't mean they are. They could be a 50-year-old man who is a big risk to them. They need to know this. We need to have some kind of class for our children run in schools about the real dangers of being online and we need to follow this up with conversations about what our kid's behaviour online is like. Why we need to restrict the usage and why they should come to us if they are approached by anyone they don't know (even if they are approached by someone they do if the communications are unusual).

If we can get them to understand the risks and feel safe about coming to us for help when they are not sure, then we will have a much safer online experience for our younger generations. If you want someone to do a talk about this to a class, don't be afraid to ask your community you will be surprised how many people will volunteer their time to help educate children about these dangers. Trust me I know a couple that give up their time for these sorts of programs now and we need more of it.

In 2019 a cybersecurity education program was announced which will be included for high school students, this is being introduced to help resolve the cyber skills shortage and help the next generations to have a better understanding of cybersecurity threats. I feel that this is not enough though (yes, it is still great) but I feel that this needs to be addressed earlier in schools. I feel that it should be discussed and addressed in the latter part of primary school so that our youngest members of the digital world are not as vulnerable to threats. We can't just leave this up to schools to deal with though we need to all do our part to better educate our children. Be involved.

Let's start a conversation about this and spread the word. Get this in front of as many parents as possible, get our governments on board to get this in schools but let's not just say it's too hard and leave our younger generations at risk. We can't do it alone but together we can make a real difference.

13

HOW DO WE GET OFF THE BACK FOOT IN SECURITY?

I don't know about the rest of you but I just seem to have this feeling of groundhog day (If you haven't seen it google it now so you know what I am talking about), I just seem to constantly be fighting battles to keep control of systems and ensure that both my clients and myself are as secure as possible. I think we are failing though, we are almost constantly on the back foot and don't seem to be making a big change or making any headwind. We just keep doing the same thing day in day out. Groundhog Day. See what I am saying here.

We keep having big breaches with a Ransomware strike that cripples three regional Victorian hospitals in 2019 and I am sure that it was not the only one of late, there were likely many others that just weren't noteworthy enough to hit mainstream media. Every day I see doctor's surgeries, lawyers, accountants and so many more getting wiped out by ransomware or similar attacks. Toll Group and Maersk come to mind.

Many of these attacks could have been prevented. Why are we failing at this so miserably?

I have some thoughts on this and I don't mind if you don't agree with me but we can get to that later on in the chapter. I believe that we are looking at security all wrong most of the time. Many people I talk to about security talk about the latest new blinky light solution or new security software that is supposed to save us all and be the silver bullet that will protect our systems from attack. Let's clear something up straight away nice and early in this chapter – THERE IS NO SILVER BULLET SOLUTION FOR SECURITY. It's that simple.

Now before all of you, vendors out there start to scream at me and tell me that your systems are awesome and will be able to stop all problems that will try to attack Blah Blah Company, just stop and think about this for a minute. Yes some of the products on the market including some blinky light solutions are really good and will help stop a lot of attacks/threats that come your way but NONE will stop everything, it's just not possible and if that is the fairy tale you are out there selling than you should be ashamed. Tell people the truth, tell them what your systems can do and be straight about it most of us prefer that approach to a pitch of promises and lies. No one needs or wants that.

Now that I have got that out of the way let's get back to it. I have talked about this next part a few times before and I am going to say it again and again until it sinks in a bit more. Forget all the new fancy solutions that we all love to play with and focus on security as a problem for a moment. If we look at many of the breaches that occurred over the last 12 months many of them were possible because systems had not been patched for a particular vulnerability and the patch may have been out for months. This is an issue that we all need to fix. Find a way to patch major vulnerabilities or if we can't find a way to protect that system from that attack method in another way if the patch is not possible (don't allow it to be accessible from the internet if you can't patch it).

When you do system upgrades, exchange or any servers, decommission the old systems. Don't leave old systems online that are not being maintained and are no longer needed. Yes I know clean up and decommissioning after an upgrade isn't exciting work but it needs to be done. Look at Australian National University, if you look at how the attack was conducted and how data was extracted. They used an old email system that should have been decommissioned but instead had been left in production. I am not rubbishing ANU IT or their Security team here, we all have demanding jobs and are constantly being pulled in new directions but we need to ensure that projects get finalised. Don't leave old systems or services that are no longer needed online, this leaves more attack vectors for malicious actors. Why would they go after your new shiny security toy if you leave the door open on an old access door at the back of your home?

ANU, since I have already brought them up, were breached by some type of phishing/social engineering attack, not just once but several times. The first attack was made externally but the continued phishing attacks were carried out using an internal email server that did not require authentication (that last part is a problem in itself) but a new email service was implemented which was more secure but the old one was not taken offline. Take old services down, reduce the attack surface. Please.

Then we have user awareness training (not specifically about ANU), many companies see user awareness training as a once a year thing that they need to do to meet compliance. That is crap, this needs to be looked at as a constant process that needs to be continuously worked on. Many attacks are gaining entry to networks via phishing or similar social engineering methods. It is suggested that this is the entry point for the Victorian

Hospitals that have all been infected with ransomware. Let's all do better at this, do it often, make it memorable and don't make it belittling as that never works.

Last but not least for this chapter, we need to ensure we minimise or reduce the risk of malicious actors spreading through our systems by limiting permissions to only what is required for users to do their daily jobs. This will not only minimise privileged access but also help reduce insider threats. If staff don't have access to sensitive content it will minimise the temptation for potential insider threats. Segment your networks into departments and don't allow area cross over that isn't completely necessary. if a breach occurs and you have adequate segmentation you could greatly reduce the level of destruction that could occur.

As I have seen indicated by many of my peers (and myself) it is not if but when an attack will occur so we need to ensure that we put up roadblocks in the malicious actor's path. Make it so hard that they just give up (or we find out they are there and kick them out).

I know I have said all of this before in pieces in different chapters but we need to do this or we have no chance of moving forward, getting off the back foot and having some wins in this cyberwar we are all fighting together (even if you don't want to collaborate, we are all still in it together).

Just to make my point clear, we need to do all the basics better. Nothing is perfect but we need to ensure that we at least don't make it easy for the attackers. Let's roll up our sleeves and get our hands dirty here. Let's do all the boring things, get them right and then do them some more. Then and only then go out and buy some new cool blinky light gadgets/platforms and go your hardest to try and come as close as you can to having perfect security (it's not possible to have perfect security but we can certainly try).

If we can all do that and maybe collaborate a bit more, I may be writing a celebratory chapter this time next year about us starting to make a big difference. Wouldn't that be great? So let's go dig around in our systems and have a spring clean out, batten down the hatches and do everything we can to get all of those basic things done.

14

HOW FAST COULD YOU RECOVER AFTER A RANSOMWARE ATTACK AND WHY YOU NEED TO KNOW?

It is 5 am you wake with a bit of a fright, your phone is ringing and as you lean over to see who it is you see "Work – Security office". It stops ringing and you see that you have had six missed calls from them. Oh no, what could have happened for them to call me with that much urgency? I am the security manager, people only try to pull me out of bed at 5 am when all hell has broken loose or a full systems catastrophe.

I have a pit forming in my throat, this is not going to be a good day. I startle with the phone jumping to life again in my hand, "Work – Security Office" displays flashing on the screen, I swipe my finger to answer the call moving the phone up to my ear as I start to make my way out of the room so I don't wake my wife.

A voice in a strange panicked tone almost shouts down the phone "you need to get into the office now, everything is locked up and nothing is working" I go to respond but the line goes dead. Things must be bad for them to hang up on their boss. I quickly get dressed and ready for the day before rushing off to the office, it's a quick 30-minute drive to the office at this time of the day which is great. I swipe my access fob at the gate for the parking garage and make my way down to my car park under the building. I park and gather my things before locking my car and heading over to the elevators. It's almost 6 am now and I know that we have about 2 hours before the staff will start arriving for the day.

Ding the elevator sounds and the doors open, I step in and select my floor, I wonder what I will see when the doors open on the 21st floor. I watch as the numbers climb, listening to that awful elevator music until it reaches level 21. I take a deep breath as the doors start to open. As I step forward I can see on all the screens in the foyer, a message "we now own your systems, to get them back contact us for a fair price"

I look around as I make my way down the hallway towards the security operations team, the same message is on almost all of the systems. I finally reach the door after what has felt like an eternity, I reach down and swipe my access fob and enter my pin to gain access to the room. One more deep breath and then I push forward.

As I enter the room I see several of my team are already here with the two-night analysts and they all turn to look at me as I approach. The look on their faces isn't good and I probably have a similar look on my face. "What is the status? Have we lost all systems?" They look at each other and one of the night staff says "more than 90% system loss, no access to any platforms. What do we do?"

This is a scenario that none of us wants to be in right? I know I never want to be in that predicament but it happens, honestly, it does. I have helped a few companies recover from ransomware incidents over my time, some worse hit than others but this is an event that can bring a company to its knees. You can have all of the best protection in the world and have the best-trained staff but mistakes happen, it only takes one.

So let's say the worse has happened like the hypothetical scenario above, you need to do a full systems recovery? How long will that take? Are your backups isolated from the rest of the network to ensure they are not encrypted along with the rest of your systems?

Do you have a hot or cold site ready to go if needed with the ability to restore your backups quickly? Could you have systems operational in two hours before the staff are set to start arriving or at least operating at a minimal capacity to allow as close to business as usual operations?

Many companies don't know the answers to these questions, the ones that do probably couldn't tell you if the backups work. Why? No one tests backups or the restore capability regularly if at all. People are busy and testing these types of things usually is in the "would like to do but don't have time" basket. Look I get that and agree daily grind can get busy but planning for the inevitable day like this in today's environment is very important. We all need to find the time to plan and test and plan some more.

How about you set yourself some simple targets and anything above that is a bonus.

1. Make sure you have a disaster recovery plan that has at minimum had a mock run through with your teams so you can iron out any possible issues.
2. At least once a quarter test a restore of your backups (at least mission-critical ones), if they don't work fix them and test again. Once a month would be better but I understand if you are time poor and that isn't an option.

3. Know how long it will take to recover your systems, use the test restores to calculate the potential time it will take to get systems operational again. People will be easier to deal with if they have a realistic time frame to work from and the guideposts don't keep changing every couple of hours or worst case days.

4. If you have a cold or hot site test the recovery at least once every 6 months or 12 months if you are short for time but make sure it will truly work as needed in the event of systems-wide wipe-out like the crypto scenario above.

5. Ensure that your backups are isolated from the rest of your networks and are not compromise-able by a malicious actor or a cryptovirus – VERY IMPORTANT.

The list above is not meant to be exhaustive and is just a few options that you need to do but the main point you need to take from this chapter if you get nothing else is – Test your backups and make sure your recovery plan works. That's it. So please put some time into this, you can thank me later if all hell breaks loose and your plans work as tested.

15

I JUST HACKED YOU WITH A DRONE

I have some red team/pentest jobs coming up and I was trying to think of how I could get into an organisation with a bit of style while having some fun. I could just walk through the door and plug in a USB charger cable somewhere or send some elaborate phishing emails that have been artistically crafted to be so perfect that even I could fall for it. No that doesn't seem that exciting to me, how about I break into your company systems using a drone? Now that does sound a bit more entertaining, doesn't it?

I could load up a special attack platform on a tricked out raspberry pi, attach it to a reliable drone with some great range and snap I could attack mobile phones, tablets, Wi-Fi networks and possibly much more if I get creative. In all honesty, I am slow to the party when it comes to using drones for hacking, there are several examples of this method being used. One particular drone concept was called Danger Drone, the creators of this unit stated that it could be essentially a hacker's laptop but could be flown to a target location with minimal visibility of the device and no real threat to a malicious actor/red teamer. Back in 2016/2017, they did several basic demonstrations of how this device could be used for flyby hacking or mobile phone attacks to help find an entry point.

A great example of the effectiveness of this unit was the wireless mouse hijack, in just a few seconds the victim's laptop was breached, the guy picks up his coffee cup to presumably go make himself a refill and he is only out of the shot for 10-15 seconds. When he gets back there is a message on his screen indicating he had been hacked. This would be a very effective attack method especially if you could boost the range so you aren't hovering just outside the window (it's a bit of a giveaway) or let's say the office was on the 20th floor, you could just scan the office for an empty desk, move in and boom breached. This attack is a little outdated but let's think about this for a moment.

Since 2016/2017 drones have shot into mainstream use with a good quality drone setting you back anywhere from a few hundred dollars to around $3-4K.

My choice for an off the shelf option would be a fishing drone, as they will be waterproof, can handle wind gusts, have a payload carry and release option (apparently for it to carry out a fishing line from the shore to extend the reach of a fisherman's cast out). They also have a good battery life as well as GPS tracking and fly home feature that could come in handy. If you're a DJI fan one of their drones would work just as well, you would just need to choose the correct model for battery life and lift ability or any other good quality DIY unit could be just as effective if you want to be a bit more custom (DJI does some very cool DIY Kits starting around $200 and I am sure there are many more from other brands as well).

Create yourself some sort of raspberry pi or mini pc setup that is light enough to not chew through the battery life and you have a very effective mobile hacking platform. Just think about what you could do with a machine that you could fly into place and within minutes be actively breaching your target networks without anyone any wiser.

Let's run a scenario here, your business is a large enterprise with offices around the world, your office in Sydney or Brisbane is on the 21st floor and Wi-Fi is available 24/7. The Wi-Fi uses WPA2 security with a router that has not been updated in years. Computers are regularly left on and unlocked. There is even some iPads and tablets that are in the office with some not having any security access requirements and Bluetooth always on.

An attacker has created a custom drone with an integrated video, Wi-Fi raspberry pie configuration that has a Wi-Fi range of 100-150 metres. The raspberry pie has a custom set of automated attack tools that it can run against multiple targets via Bluetooth or through Wi-Fi once the network is breached. Within minutes a drone of this type could gain access to your systems install malicious or remote access and be gone.

How do we defend against attacks like this? It's a tuff vector to defend against but there are some things we can do. Systems such as Geo-fencing, radar can be used to help detect and stop drones from getting too close to company assets. Geo-fencing is a method used by airports to stop drones from entering their space but there has been evidence that the manufacturer configurations on drones can be bypassed (and have been) but the best protection is good old fashioned security hygiene.

keep systems up to date so vulnerabilities are not available, make sure Wi-Fi is configured to be as secure as possible with all routers patched with the latest protections. Ensure machines are locked or turned off when staff are not in the office, use multifactor authentication and if you leave devices on making sure they are locked. Simple and very effective solutions to a growing security threat.

Look I think this idea is pretty fun and may build myself a prototype one day but at this time I think I will stick to the good old (on land) security attack methods for my red team engagements. If you have already built yourself a hacking drone or similar I would love to hear about it? Or maybe you have some thoughts on how we better protect IoT infrastructure from these types of attacks moving forward.

16

IMPOSTER SYNDROME A REAL ROADBLOCK FOR CYBERSECURITY

Over the last 6 months or more I have been having quite a few conversations with my peers in cybersecurity and they seem to all include a form of what I believe is "Imposter Syndrome". What I mean is that many conversations start with my articles for CSO or my first book "A Hacker I Am", how that person likes them and asking me how I started my writing journey. The conversation generally takes a bit of a turn at this point and I want to discuss this, really get in and find out how we can get past this problem.

What I am talking about is imposter Syndrome. It's a strange psychological issue that most of us have in some form or another, it's that self-doubt that creeps into our subconscious and tells us "you don't have something worthwhile to say", "how did you get this job, you aren't qualified enough", "no one will read my book or chapters, why would what I have to say matter" and I could go on for pages and pages of these. This is not something that is an isolated issue, it is much more common than you think. CEO's, CISO's, brilliant public speakers, managers can have this same problem? Seriously reach out to a mentor or someone you look up to in your field many of them will indicate that they have experienced or still experience imposter syndrome type self-doubts.

I was reading a chapter from Psychology today "The Reality of Imposter Syndrome" from late last year and as I was reading through the chapter, I had a realisation that wow I have had these traits my entire life. In the chapter, it refers to imposter syndrome as a "psychological term referring to a pattern of behaviour where people doubt their accomplishments and have a persistent, often internalized fear of being exposed as a fraud". The term was created by clinical psychologists Pauline Clance and Suzanne Imes in 1978 when they found people who had imposter syndrome, even with a lot of external evidence of their accomplishments they would still be convinced they didn't deserve their success and would just push it aside dismissing the success to luck or something else.

I found myself doing something along those lines at the recent Cybercon in Melbourne. I was having a conversation with two very well-known individuals in the cybersecurity industry. When the topic crossed to me and it was stated that I was one of a few superstars in the Australian security industry. My immediate reaction was to dismiss the statement, I responded that I was just a security guy trying to play my part in the industry and that I wasn't a superstar. Look I still don't think I am a superstar, I think we are just all here to do our part. Yes, I work very hard to share my opinions with both my CSO articles and my books so that I can make a small difference.

Yes I know right that sounds exactly like imposter syndrome, doesn't it? I think in a way I do have a self-confidence issue that can come out from time to time but I am in my opinion I am one of the lucky ones, as I push past those doubts and keep moving forward.

I don't let those little internal whispers hinder me. Sometimes admittedly I probably should have listened to them but it is my opinion that if you don't take any chances in life and try to achieve your goals then what are we even doing. We need to ignore those thoughts and if you have a burning desire to share your opinions or suggestions that you think could make a difference DO IT...

Write chapters, write a book, start a blog, get up on stage at any number of events (Public speaking is my kryptonite, I find it very hard to do but I still push myself to do it, oh and there is the hole COVID-19 situation we have to deal with before any of us have to worry about that) and as you do it more and more it becomes easier. Look I am not saying that it will be all smooth sailing and whatever you do will be well received.

There is still a chance people won't like whatever it is you do but how do you know if you don't try at all. I see failures as lessons, in self-insight. Self-insight is never a bad thing if you ask me, so what if you are not cut out to write a book, maybe you are destined to be an awesome speaker. Don't look at the things you can't do as failures just think of them as something more you know about yourself, then focus on what you can do.

Strive to be better and do your part in whatever industry you are passionate about. if we are lucky you are passionate about cybersecurity and your next idea may be the one we need to turn the tides of the constant battle we are all waging to protect our ever-expanding environments.

17

IS 5G OUR NEXT BIG THREAT OR AT LEAST THE BRIDGE IT WILL RIDE IN ON?

All I keep seeing is articles and talks about 5G being rolled out around the world. Australia is pushing ahead to try and be one of the leaders in that market. I honestly understand why, the preliminary tests of Australia on the Telstra network which has been rolled out in five cities and a test at their Sydney CBD office achieved a download speed of 1.216Gbps on a new Samsung Galaxy S10 5G phone.

Now these results are a little misleading as the network isn't being used at the time of the test and you will not likely get those types of speeds in real-world scenarios but I would assume that at a minimum you could expect to have speeds of 2-3 times that of 4G. That would mean that you can expect to get easily similar speeds to the top tier service on NBN (100Mbps).

Australia is the third country to adopt the 5G technology which has had some controversy due to the ban of Huawei technology being used as part of the core systems. It may put some challenges forward for the NBN as it will be a very comparable speed but it will come down to what price point will be offered for data (currently listed at $80 for 100GB and $100 for 150GB still quite expensive but maybe a viable option).

If 5G is available on great data plans it could be worth a look but it will have limited availability for quite some time until the rollout has been completed across more regions. I will be sticking to my NBN FTC connection for quite some time (although I was on a 3G connection for about 2 years waiting for it to arrive- it is pretty good though) but this rollout will bring with it great opportunities.

IoT and Autonomous vehicles will need this network to be able to truly communicate in real-time with all of their different systems to ensure that up-to-date information is always available to all systems that need them. In autonomous vehicles, this could be used to know what traffic is like along its travel path and adapt its routes or even react to pending danger but none of this will truly be possible with a network that can handle the bandwidth pressures that it will bring with it. Look speed on 5G will be great and with self-piloted drones and those autonomous cars, we will have a very entertaining future, especially in security when we are tasked with protecting it all.

That is something that I haven't seen discussed much which concerns me a little, 5G as I have already indicated is going to be quite fast and will likely improve over time as technology is developed, so we will see millions of IoT devices suddenly start to be rolled out onto our cities, homes, cars or almost everything but currently security is only an afterthought for these devices. I know there are some reasons for the slow introduction of security in IoT devices as I completed a research topic on this for my last unit of my master's degree. Some issues specifically are to do with the actual power and processing capability of these devices or lack of. They just can't run it without also doing the job they are made to do. These devices need to be built in the first instance with security in mind but this may need some technological advancements before it will be a true reality.

Don't get me wrong I think security on IoT needs to be a priority and believe that it needs to be a the forefront of all design concepts through to final products no matter the size or cost issues it brings with it. It's just something that needs to be done. If however security is not improved (or introduced in the first instance) then this awesome 5G network that will put us at the forefront will be the bridge that allows our systems to be attacked and controlled by malicious actors. Think about it, these devices will be connected to a fast network with basically no security and will likely sit inside our critical infrastructure, in our cars and our homes.

I think this is an issue that needs our attention and fast before it is too late and APT's or malicious actors have unrestricted access. These devices need to be monitored and encrypted to help reduce the ability for the data to be used against us, we need to develop the ability through new processes or systems to ensure that we can monitor them in real-time. If we can't it will be too late to do anything about it. With those speeds, any attacks could be completed quickly with minimal threat to the malicious actors being caught in the act.

What about the mesh type grids that will be relayed in real-time to some of the passenger drones so they know where each other is if you could manipulate that image the damage you could inflict. Large files could easily be transferred back and forth between attacker and victims with almost no chance of detection at this time. That's a really scary thought don't you think, currently we have minimal control on our networks (yes we all try to have total control but in reality, we have known we don't have full control) and with speeds expected to go through the roof and millions of insecure devices being added we are going to lose control it's inevitable so how do we fix it?

Honestly, I don't know how we can fix it, not completely but I believe that we need to stop and go back to the beginning with implementing real security aligned IoT devices, stop the avalanche of insecure equipment. That's a good first step, then maybe regulate the IoT industry (don't all start screaming at me now and let me explain). I do not like red tape and regulating things but in my honest opinion, I think it will be needed to enforce a standard of security that will be needed to have any chance of securing the billions of new devices that will be introduced. We need to look at this as a public safety initiative because, in the end, it could be the factor that saves lives.

Alternatively, we could leave things on the current path and wait for "It" to happen, a complete disastrous event that I am not sure what it will be at this point but whatever it will be, will show us that we should have acted sooner. So let's not sit back on our laurels and do nothing, let's come together and do something for the better of our communities. Our children will likely thank us later for it.

18

IS MFA OBSOLETE BEFORE MANY EVEN ADOPT IT?

Multifactor authentication or MFA for short. I have written about MFA before in which I told you all that you need to stop resisting MFA because so many organisations are still not adopting it, "it's too hard, I don't have the budget for that" you have probably heard the same comments as I have. I still recommend everyone use MFA for everything, yes some methods can be used to bypass it but that doesn't mean you shouldn't implement it.

Let's look at some of these methods that can be used to bypass 2-factor authentication. I watched Kevin Mitnick do just that onstage at this year's (2019)

Cybercon in Melbourne, he hacked his google account onstage, look this isn't something new and the proposed attack method has been around for about a year but it has shot into the mainstream more recently as the method is starting to be utilised by the malicious actors. As Kevin explains in his video the attack follows this sequence. The malicious actor, in this case, Kevin sends a phishing email to the proposed victim, it will follow the usual process of bringing up a page that requests your login details.

You as the victim would enter them unsuspecting that it is a fake site. In the background, the malicious actor's systems would have an automated process that immediately logs in with those new details and then redirects you to a page requesting the 2FA (MFA) code, which is sent to your app or mobile device that you enter in the page. In the background the malicious actor's platform will log in with your provided code and woo Lah they are in. you would then be redirected to the real site and you will just try to login again thinking it failed somehow.

This method can work on many platforms not just google, here is an example of Microsoft office365 and there are many more online platforms that are susceptible to this attack. Okay so now some of you will be reading this and be saying "well what is the point of implementing it to start with if you can just bypass it". I know some of you will be doing this, as while I was making my way out of the auditorium after the presentation by Kevin Mitnick at Cybercon I could hear several people saying just that.

Now before I continue I want to say something about that last sentence, what the hell? This was a security conference and a couple of the people I heard say "then why even use MFA?" these were people who are tasked with protecting companies from threats. Wake up just because there is a possible way to bypass MFA doesn't mean you shouldn't use it. We are supposed to be the professionals here, start acting like it, please.

Okay, so mini-rant finished. Back to MFA. Look yes, I know MFA is not perfect and in some instances, it can be bypassed but a couple of sayings I have heard quite a bit come to mind "don't put all your eggs in one basket" or "Hedge your bets". Let me explain what I am getting at here. MFA is not the only protection you should have in place, this should just be a piece of the puzzle.

Yes have MFA with either google authenticator or another providers authentication app there are a few to choose from, combine it with a strong password (even use a password manager if you like – Lastpass, Dashlane, 1Password) and obviously don't use the same password for all of your accounts this is a sure way to get your account breached. I am a fan of password managers, especially for non-technical users because you can have a really strong unique password automatically generated for each different account and you don't need to remember them. You just have one really strong master password and setup MFA so that it requires a secondary push app authentication or even some sort of hardware authenticator like Yubikey. This will keep things simple and secure. Not much messing around but great security. This is just my opinion and I know some people don't like password managers but I think it achieves a great result with reduced risk compared to not having it.

A secondary bonus about using a password manager is that in the described situation above, the password manager will not auto-load the login information because it isn't the real site, the PM will see the slight variation of the domain that we humans might miss. In some cases that may be enough to make a user pause and say hang on something isn't right here. Doesn't that make the PM worth a second thought?

Let's not stop here though, as we should all be implementing click protection for email services to help block users from clicking on dodgy links, it's a great protection that yes may not protect us from everything but this is all about stacking the odds in our favour, don't you think? Put in great endpoint protection, use click protection, implement the best MFA you can get and then train your users so that they can recognise things that don't look right. That is not an exhaustive list here and there are more things we can do and yes many of these things on their own won't stop an attack, very true but put all of these things together and you are going to start to make the malicious actors job very hard.

If you can make it hard enough so the effort required to get in is much higher than the fruits of their labour then they will just go attack some lower hanging fruit. Trust me there is a lot of it for them to target (all the ones who still haven't implemented any MFA would be a good place for them to start or the millions of accounts who still have the same passwords that were in breaches from years ago). Segment your networks, isolate different departments and put firewalls between them. The list goes on and honestly, you will never do all of them but please do all of us a favour and at least start with two-factor authentication with a good password.

So, yes you are right you can bypass MFA but remember what I have said it's about stacking the odds in your favour. Nothing is concrete and infallible but that doesn't mean if we put it all together we can't get close. If it makes it more fun for you to make it a game, why not think of it as putting obstacles in front of malicious actors just to get under their skin. Confuse them, have some fun with it and don't make it easy for them (where is the fun in that).

19

SECURITY AWARENESS TRAINING: HOW I APPROACH IT

Security awareness training is something that is discussed quite a lot these days and it is commonly something that is just completed to ensure a corporation meets its compliance, but this is not the way we should look at security awareness training. I believe that it is something that we need to do regularly to help educate and share knowledge on security with all members of our organisation. It is not about learning all about security and all the tools we use or any of that, it's about making everyone we have interactions with safer online in their personal lives and better-protected users at work. It's a noble cause that deserves our utmost effort but in my opinion, is badly done in so many instances.

Now, don't get me wrong I am not saying that my way is right and yours is wrong. Every method has merits and we need to look at what works for our organisation and what doesn't. There is no point doing a rigid formal training program if that doesn't fit your company culture or style. We need to truly consider what will work and not just do a quick google search and then buy the latest buzz version or style of aware-ness training to make sure we cover our compliance requirements.

Do you know what staff do with those online secu-rity awareness training programs? They start the video and then continue with their work, they click next when required and honestly don't learn anything from your probably expensive training program. So why not try to do something a little different, actually try and teach your staff something worthwhile. Maybe even something that will help them for the rest of their lives.

Look I know it's hard and we are all strapped for time so do what you can but make every minute you have available worth it. Do some face to face training, don't do boring, no one will remember boring. Personally, when I am conducting training programs I try to make it a bit humorous so that attendees remember what I have tried to tell them. I deliberately keep it simple, this isn't because the attendees are stupid (many of the people I train are lawyers, accountants, doctors and just all round smart people) it's because that's not what they do, they don't understand our language so why try to speak tech when they understand English or if you know how they talk (their tech language) why not bring some of that into your presentation. It will make more sense to them and be a better use of their time and your own.

A regular saying, I use in my presentations is the good old "passwords are like toothbrushes, pick a good one and never share it with anyone", have some fun with it you don't need to put everyone to sleep. I often play a bit of a game if I am talking and I see people start to drift away. I will get the group to try and come up with the silliest four-word passphrases they can. You need to have some fun with it if you do the attendees will respond. I have heard some very creative options over my time, it can get the group involved and attentive to what you are saying (which is the goal, so they remember).

My attendees always seem to like it when I tell them that we have been making users have difficult passwords which they can't remember but are easy to guess for computers just because we thought it was the best thing to do. Most complex passwords can be cracked in a relatively short amount of time, but a good passphrase will take years and years. We don't even need to change them as often as we do (that's another thing we added just to make things more difficult) as long as we keep them secret (Which is easier said than done but doable). I am sure you can see what angle I am going here, have a bit of fun and even poke a bit of fun at ourselves if it will help the attendees enjoy themselves as well as take some knowledge away from the session that can help them be more secure.

So, the training session is done, and you mark it done for another year. No need to worry about training again for at least another year, right? Wrong. You need to do more than one training session to make a difference in your users level of security awareness. Quarterly sessions would be perfect if your organisation is of a size that you can get this organised or even if that one session is all you can get in a face to face group training follow it up with some funny emails every month on just a quick topic, have some funny posters made up about financial scammers or some sort of phishing email that helps identify malicious emails. If the users see them regularly they may remember them when the time comes, and a real threat arrives in their inbox.

What about adding a screen saver to all machines with a set of crafted slides that reminds them to lock their computers when they leave their desks, not to share passwords and like the flyer above help them identify threats via a simple graphic that displays tell tail signs. These are just some ideas that I have used in my training but there are so many more options that could work we just need to think outside the box and come up with ideas that will work. They don't need to cost a lot of money to be effective but we just need to not leave it with the single basic online course (hey, by all means, do that as well if you can – just find something funny so they will actually watch it) or the single physical training session (I know some people that knock these out of the park though).

You need to do more and make it an active part of your security program, humans are in the end a big part of how we will, in the end, win the cyber battle. I have heard some of my peers say on numerous occasions that this is not a technology problem but a human problem. So yes, let's do what we do best and get all of our security systems in place but don't leave all the humans out in the weather, stand together as an organisation as one and help create an educated united front.

20

SECURITY COMPLIANCE IS JUST THE START, NOT THE FINISH LINE

So, your company was certified for ISO27001, PCI DSS or whichever standard you are either required to align with because of your industry or legal obligations and you think your security job is finished. Hahaha, not even close. Just because you have jumped through the hoops and fought through the red tape to get that pretty certificate and snazzy logo to display on your website or in your foyer to say you have met the requirements doesn't mean that you will not have a security breach. If you think it does then please do both of us a favour and read this chapter a couple of times, so I can convince you otherwise.

Compliance should be considered as just the first step in a continuous effort to better your security standing, Security compliance is by no means the finish line in your race to protect your organisation. The sad thing is that many organisations only complete these certification processes for publicity or reputational benefits, I am told time after time by companies "we have completed our compliance requirements this year we don't need to worry about security anymore", please don't fall into this trap. Let me be very clear here, I am a certified ISO27001 auditor and I think that aligning your organisation with a framework such as this is a great idea.

If your organisation aligns with a framework and works towards becoming certified then it can only help you improve your internal policies and procedures which is great. I am sure that you will improve the actual security of your systems with the introduction of formal processes. I believe that there are many benefits to this process and completely recommend it (just wanted to be clear on that). The problem I want to focus on however as you would have gathered already is stopping here, thinking that certification is the finish line.

In the first month alone of 2019 there were many big-name breaches in Australia such as Victorian Public

Servants, Nova Entertainment, Hawthorn Football Club, Big W, Early Warning Network, Victorian Government, Department of Planning and Environment NSW, First National Real Estate, Fisheries Queensland, Optus. Now there is more but I thought the first 10 I found would be enough to make my point here, all of these companies would be aligned with security governance/compliance frameworks and believe would be either certified or working towards certification under more than one of them. However, they were all breached in January 2019. The reasons why are could be many factors, but I don't want to dive into each scenario and determine the specific cause, but I want to alternatively draw attention back to the fact that certification does not mean you will not have a data breach.

If we truly want to start to make a difference in this war that is being waged in cyberspace we need to understand that fact, Yes go out do everything we can to meet your compliance requirements, yes get certified. I encourage all of you to do that. Once that is done though (although compliance certification is never really done), go back to basics on security. Train your staff (not just once to get the compliance tick), set up a plan and try to improve on awareness constantly. Any improvement will be a benefit to the organisation, most breaches are phishing, or social engineering type attacks these days.

So, train, train and train. Make it fun, make it memorable but do it and try to do it well. Work on network segregation, make sure you can't reach backups from your primary network, separate departments. Then make sure that your systems are updated as much as possible (I know some of you will still say you can't but update what you can and what you can't make a plan so that you can). I just want you all to not just think compliance means that you should stop trying to improve your security, it is just that first important step in many that we all need to work towards, but it is just that, One step of many.

I know you all understand what I am trying to get at here, let's not treat compliance as the goal but just gain compliance as part of our overall security improvement process. If we all do this and try to do the basics right then maybe, just maybe we can make a difference in this battle that is continuously testing our skills and resilience. We all get it, we know the challenges we each face on a day to day basis, but together we can do it.

21

SELF-PILOTED PASSENGER DRONES COMING TO AUSTRALIA SOON BUT WHAT ARE THE RISKS?

I am going, to be honest, here, I love the idea of passenger drones and flying cars. I grew up watching "The Jetsons" (I just watched the intro, it's crazy how bad the graphics were back then but it brought back some memories) on Saturday morning's thinking that would be us one day. It's funny when you think about it, some of the engineers who are putting these vehicles together today would have been growing up thinking the same thing as me. Long have movies, TV been a place where brilliant people see something and say "you know what I can make that for real" and they do.

Passenger drones are not quite here yet though which I am admittedly very disappointed about (which I am sure you would all have guessed) but honestly they are becoming reality. Especially if you are talking about passenger drones. A report by Deloitte "Elevating the future of mobility" from January last year is an interesting read and discussing some of the challenges that the industry faces. The report/chapter predicts that we could see Passenger drones as early as 2020, some form of traditional flying cars between 2020-2022 and revolutionary vehicles could become a reality by 2025. As Deloitte indicated these vehicles will have some big regulatory and safety hurdles to clear before they will be able to become a true reality but testing has already begun,

A half-plane half drone nicknamed "Cora" has already been secretly been testing in New Zealand, it can carry two passengers for a distance of up to 100KM at a top speed of around 18 kilometres per hour. It has vertical take-off and landing but works much like a plane once in the air with one rear propeller pushing it forward. The company expects that they will have "Cora" passenger drones as taxi's over Los Angeles by 2020.

In Australia, we are destined for an Uber Air service to start in Melbourne by 2023 (some predict it could be as early as 2023. This service has been indicated to be a short airport to CBD service initially (19km) that can take 25-60 minutes in a car will take 10 minutes in the new services. That is a massive reduction in time especially in those peak times, I guess its success will come down to costs. In my opinion, if they can provide these trips for $100 or less I think just the convenience alone would make them viable and honestly, I would do it at least once just for the experience factor. In reality though if these services can be provided long term at a price very comparable to that of normal Uber trips (obviously at a slight premium) this could be an extremely successful operation.

Uber is not alone several more operators are entering the market (or at least planning to) such as Ehang 184, Airbus Vahana, Boeing PAV, Bell Nexus, Intel Volocopter E-volo, SureFly just to mention a few. The idea of removing a 1-hour commute and replacing it with a quick 10-minute drone flight to the office sound awesome to probably millions of people living in cities around the world but have we considered all of the risks? I am not talking about the risks of crashing or systems failure but from my perspective, I am thinking about cybersecurity threats.

If you look at many of these passenger drones they are heavily controlled by technology and are reliant on a lot of information from their internal sensors/ technology but will rely on a lot of external systems as well such as 4G/5G technology, radar and many talk via a mesh framework that will allow them to know where other drones are to navigate each other with ease. External access like this brings great risks that I have not yet seen covered. What happens if a malicious actor hacks the mesh network? Just imagine what they could do, remove objects from the systems path so you have drones running into buildings or other drones (one way to assassinate someone), what about if they gained control of these devices (most are autonomous) so they could then use them to destroy targets like communications, critical infrastructure or even be used by terrorists to send dozens of these into a crowded place to inflict harm against modern society. If you think about it there is a lot of possibilities but what can we do to better protect these passenger drones as they start to become a reality.

The first step is quite simple, make sure that all of these systems can be manually controlled by passengers not fully autonomous, secondly, and think of security in the first instance. Don't try to add security to these systems later, bake them in from the start, and truly test them (I am sure there will be lots of security professionals/Hackers who would love to test these systems for you). How about ensuring all communications are encrypted (seems like a bit of a no brainer), at minimum, this will ensure we have some control.

It's going to be a big challenge to ensure that these autonomous systems are well protected for both autonomous cars and passenger drones but we need to do this to ensure we can keep ourselves safe. We all know that no systems are 100% safe from cyber-attack, that's just the way the world is but that doesn't mean we shouldn't do everything we can to make them stronger. 5G will be one advancement that will add further challenges, the speed of connectivity will be required to enable these devices to work well but that speed also allows for fast methods of attack. Then we will have cheap IoT devices that will be used by these systems to allow them to gather further information about their surroundings. These devices will not have any (or minimal) protections leaving the wholes systems at risk.

We are certainly going to have a challenging future with all things interconnected but we can't mess around with things like passenger drones or autonomous cars the costs if we do will be too great. So let's start to come together as an industry or community as a whole and make sure that these protections are in place to keep us all safe. Oh and if any of my readers are working with one of the many drone start-ups I would love to come and have a test flight you just tell me where ◈.

22

STOP RESISTING MULTIFACTOR AUTHENTICATION

Over the first 6 months of 2019, I am seeing more and more accounts being taken over by malicious actors on office 365 and google mail where the attacker will create some rules in the backend to redirect emails of interest into deleted items, RSS feeds or any other folder they think may be overlooked by the user. They would then either use the account and information gleaned from the account or redirected emails to try to get clients/customers of the business that the account belongs to make payments into a new account for outstanding invoices or something along those lines.

The malicious actor would then use the accounts to spread their access to further accounts to continue the cycle over and over. I see this type of scenario all the time and the sad part about this is that there is a simple solution that may protect many of these victims from having their accounts breached in the first place. MULTIFACTOR AUTHENTICATION. It's pretty simple and on most platforms is turned on very easily with no additional costs. In office 365 to turn it in for the organisation, it is as easy as selecting a tick box. Then all users in that organisation can log in to the online portal and run through a very simple wizard to get the two-factor setup. You can use a text or the authenticator app (the app is the best option, but the text option is still better than not using it).

Look, yes I get it I am sure that the change over will not be perfectly smooth and you could have some interruption by doing so with some users who have trouble with the process but honestly, this small amount of pain will be much better than explaining to the CFO or CEO why you hadn't done it after an attack could have been stopped and prevented a $50K payment or worse to a malicious actor. Come on you know I am right!

I am sure that google mail will have a similar easy process that can be carried out to allow for the same two-factor authentication to be switched on and used. So why do I keep seeing many businesses that don't have this turned on? It just seems stupid to me, I just honestly don't get why organisations don't turn it on. Is it our fault (IT, Security industry) for not getting the message out there for why it is important? Or is it because businesses just don't know what they should be doing in the first place or even that it Is an option that they can even have in the first place.

I think it is a little bit of all of them, we as the protectors should do more to help businesses protect themselves, we need to look past the money and actually help them get the basics right in the first place. The funny thing is if you do put the business first and help them in knowing what they need to do and not just push to sell some more blinky lights then they may just become your best customer you have ever had because of that trust you have created. You won't just be about the next quick buck you can make from them as many businesses think of the MSP or MSSP industry. This goes for the whole industry though, do the right thing by people and they will remember.

So, if it isn't all our fault (it really isn't but we can do better), what is the problem that stops users and organisations alike from adopting MFA (Multifactor Authentication). I think that it could be a couple of reasons and likely different for each user or business. Laziness is certainly one of the reasons, people like to do things the easiest way possible, the path of least resistance. If all they need to have is a password then that is all they will have. Yes, they might know that they should do MFA with either a text or app but if they don't have to do it they will just do what is the easiest option. We all do it sometimes, admit it we all do. This fact alone is why Shadow IT was born, they needed something and couldn't get it done easily so they plugged their own Wi-Fi device in.

What we need to do is not make it optional at all on the very top level, Microsoft and Google and every other platform provider don't let users access systems without MFA of some form. Simple and very effective result don't you think? I know that is easier said than done but this mandatory change will make systems much more secure and not rely entirely on passwords alone for authentication.

Now why we wait for the commotion to settle down and people to stop freaking out about what I just said you all know what I have said is right. Take away the choice and the problem becomes a bit simpler to resolve. Look I know that MFA isn't perfect and some techniques can get around this as well, but it is really difficult and is rarely successful. So just because it's not perfect doesn't mean we shouldn't use it as part of our defence. Improvements are just that improvements.

If any of my readers work for Microsoft or Google and has access to anyone who could put this suggestion to the right people, it would be greatly appreciated. I am sure in time we will all come to see that it was for the best even if it causes us a little bit of pain because of it.

Now I don't think that will happen anytime soon but if it does it will be terrific. However, until then we need to do what we can to help people understand that the option is there and that honestly, it won't be that inconvenient to them to use the MFA component. They won't have to pay any more money, but it will make them much more secure against the many threats that face them and their online presence. If we can do that I will be much happier, and I have a feeling that many of you will be as well.

Let's do this together and reap the rewards with a better-protected world. As always tell me what you think. Disagree, I don't mind. We always need to look at other opinions and options. If we can't do that we will never stand a chance in the virtual war, we are fighting.

23

THE RIDESHARE AND PUBLIC TRANSPORT HIDDEN THREATS

I don't know about the rest of you, but I seem to spend a bit of time lately in either rideshare like Uber, Didi, Ola or any of the other new entrants to the market. I also regularly travel on the train and I want to discuss some possible threats that I don't feel many of us consider when we travel using rideshare or public transport. The risks will be cyber-related not physical threats as that is my focus of expertise and honestly why you are all here reading this chapter in the first instance. I will likely go down the scenario path so that I can demonstrate these risks so we can all better protect ourselves while using these types of services in the future.

Let's take a look at public transport first as I believe it poses the biggest risks. How about we go with a very common scenario for many readers on their standard commute each day via train and bus. You arrive nice and early at the station for your hour-plus commute to work (that is the average for most cities in Australia), I know some colleagues who do a two-hour commute each way (wow that's long). Back to the scenario building, you swipe your go/opal card (depends on the one used where you live) or even pay Wave or apple pay, google pay (transit cards are being phased out in favour for these other methods). Whichever payment method you use you swipe on the train and swipe off at the other end.

You make your way on the train or bus and you find a seat (if there is one even available during peak times), you unpack your laptop or tablet and you start to do some emails or work to get a head start while on your commute. You most likely connect to the free Wi-Fi and think nothing of it. The emails you are sending could contain privileged information and could be damaging if they made it into your competitors' hands. You make calls and discuss similar sensitive information without a second thought. You arrive at the station, swipe your payment method and walk a few minutes down the road to your office. That sounds like a pretty normal commute for most, wouldn't you say? Is that something you do or witness most days? It is for me, I see this type of behaviour all the time when on public transport but is there anything wrong with that? I think there is, and I feel you are putting yourself and your company at risk without even knowing it.

Let's go back and look over what happened during the scenario. First, you swiped your credit card or pay wave (whichever payment app you use). If I was a malicious actor I could be waiting and as people are swiping their cards or tapping the app I could be charging you a secondary charge without you being aware, let's say $1 per person with between 10-20% of the community using the services.

Brisbane has 2.28 million residents if 10% of those catch public transport (228,000) and say 20% (45,600) of those goes through central station (this is just an apotheosis, not exact figures). If you could scan and take payment for 20% of them by just sitting near the scanners with a device that has an extended range, a malicious actor – I in this scenario could steal $9120 from unsuspecting victims. You could keep changing stations, fly between states and make a killing out of this. Look this type of attack is unlikely as you would still at best need to be a few metres away and it would get pretty obvious if you just sit or linger around payment gateways. I am certain after about an hour you would be having a conversation with security.

I think you need to be aware of the threat though, as the credit card payments become the standard skimming devices will certainly get made to help malicious actors steal the info or double charge you as you get on public transport. A malicious actor could also just walk through a train or bus letting an auto payment machine receiving payments as they pass you sitting down or sliding past you to get to the other end of the carriage. Buy yourself a couple of cheap RFID shielding sleeves (they cost about $1 each), that will certainly help save you from that walk by attack at least.

The RFID attack is not your biggest threat while on public transport though when you found your seat on the train you connected your device to the free/public Wi-Fi. Please don't do this unless it is necessary, especially if you are using a company device with sensitive data on it. As a malicious actor you could use a man in the middle technique to trick everyone on the train into thinking that your wi-fi network that they just connected to thinking it was the trains free Wi-Fi, they could then inspect and modify any communications travelling across the network. If you are connecting to resources without a VPN you are putting yourself at risk.

The malicious actor could even infect your machine by replacing a clean file with a modified version that was captured then re-sent on to you being none the wiser. This type of wi-fi attack is quite common and it is recommended if you need to use the internet while on public transport use your own internet or mobile hotspot. It is much safer and less likely to be intercepted (still possible though if they use the same technique on your Wi-Fi signal but unlikely unless you were being specifically targeted).

You should also consider who is sitting behind you and what they can see on your screens or what they can hear during your conversations. Any one of the people could be from a competitor, be a malicious actor or anything you don't know. So, take some advice don't work on sensitive documents on public transport or take sensitive phone calls. It isn't the best idea and honestly, it is quite annoying when someone near you is taking a loud call on the phone while commuting oblivious to their surroundings talking about things that are not appropriate at that time. Just stop and think about what you are doing.

Rideshares, do you think about the risks with these services? A payment skimming device could be installed behind a seat in the vehicle and when you get in you could be charged an extra $50 for your troubles, this could be done without the car owners knowledge, just slip a cheap device down the back of the seat and it could scan away till the battery runs flat, this could take days. That could put hundreds of riders at risk without being aware of what happened. Phone calls in this space should also be the same as on other public transport methods, don't have sensitive conversations. Let your phone go to voicemail for 20 minutes, I am sure your world won't burn down over that time and if it does 20 minutes would not likely have saved you from that fate anyway.

There are more threats with rideshare and public transport, but I think you understand now where I am going with this. Be more considerate of where you are, don't put yourself at undue risks and just minimise the attack vector. Use the time to read a book, listen to music or a podcast. It will be more relaxing and much safer for you and the organisation you work for.

24

VOICEMAIL PHISHING & FRIENDS

As many of you would probably already know I work in the cybersecurity industry and this time of year we get hammered literally. It is like it gets to mid-November and the malicious actors come together for their year shindig to decide what terrible, mischievous things they are going to get up to while the rest of the world (including us security people – well that's what we hope anyway) are out enjoying the holiday season. It's this time of year that we get an avalanche of emails asking for quick money transfers, new voicemail messages and any other new scam they come up with I am guessing over a couple of drinks and a big feast that our scammed money pays for.

Cybercrime (which is just essentially just crime on a digital platform) is big business in many parts of the world and it is very successful. So, these malicious cybercriminals have gathered, they have planned, they have feasted and now it's December 1st. Hold onto your hats, things are about to get a little bumpy. Your inbox is going to be swamped with emails both legitimate and scams, but you are going to be so busy you will barely notice that you are about to be duped. Look I get it, this time of year is ridiculous, people want everything finished now before everyone leaves for the festive break and we all get a bit swamped. We cut corners and do what is necessary to get things done. Fast.

Please don't do this, check your emails, make sure you follow the correct checks and procedures that are normal practice don't make a mistake that will make your festive season one, to remember for all the wrong reasons. Let's look at a scenario or two of what you might see and what you should do with them.

Scenario one – you get an email from a regular supplier asking you to pay an invoice but the bsb and account numbers have changed from the ones you have on file. Stop. Do not cut corners and just change the account and make the payment. Call the client directly (don't reply to the email) and ask them to confirm that the account details have changed. If they are a little surprised by this, you should advise them they should get the email account checked as it has probably been breached.

Scenario Two – you get an email request from your bosses, boss (via an email that isn't a company one or even in the company address book) asking you to do a quick funds transfer for a deal, so it doesn't fall through. Don't do it. Pick up the phone and call them. If you can't reach them inform your security or IT team and get them to check it out for you. Notify your supervisor, indicate what you have done and that you will not be completing the money transfer until you can confirm the details/identity of the persons requesting it. I know this can be scary, no one wants to be fired for not doing what they are asked.

You are more likely to get fired if you lose the company $50k or more than if you don't transfer and it was a legitimate request. If you explain the reasoning and the fact that you had followed the correct company procedures, I am certain your managers will be grateful you followed the correct procedures.

Scenario Three – an email saying urgent voicemail message comes through on the 24th December at 11:45 am and you are finishing at noon for the day. The company is shutting up early for the break. Stop and think for a moment. You don't ever get voicemail messages to come through via email, you have a flashing light on the phone, **this is a scam**. Don't click on the links, just delete or send them through to IT to look at for you.

These three scenarios are just common ones I see all the time and I thought they would help you keep a lookout for them in the lead up to the festive season.

Some key take always from this is – Stop and think before doing anything, ensure you follow correct procedures and if you are unsure contact your support team. Ask them if it is legitimate, yes, they will want to leave as well but I am sure they would want to leave for the holidays 10 minutes late than have to spend most of the break instead rebuilding the entire systems after a cryptovirus gets in.

25

WOULD YOU QUERY SOMEONE IF THEY WERE IN YOUR OFFICE ON A COMPUTER AND YOU DIDN'T KNOW THEM?

I got the idea for this chapter during a presentation by Paula Januszkiewicz during the last day of Cybercon 2019. During the presentation, Paula was telling a story of how she had done a physical security test at one of her clients at some point and she discussed how she was able to gain access to the site.

The story kind of went like this, she walked into the lobby area behind others entering the building and she saw one of them click on level 6 on the electronic floor selection pad. That was the floor she needed to get to. She indicated that she was pretty stoked by the fact that it was a pretty handsome looking guy that was going to be her mark (she said it's much easier to do what she does like a pretty girl – she is probably right). When the door of the elevator opened she immediately entered the elevator that he was going to be entering and headed directly to the back of the elevator. During the presentation, she explained why she walked to the back as having the ability to be out of sight and also to enable her to watch what he does with entering pins or biometrics or anything else but also so that he was not able to watch her too much as it would be a little weird and awkward if he did.

When he entered he looked over at her and she gave him a slight smile and brushed her hair out of her face a little shyly. He turned around and faced the door with her behind him. After a few moments he turned and she did a sort of flirty "Hi" to him and he responded in turn with a massive smile on his face. At that point, she told the audience that she knew she had him. They rode silently almost to the sixth floor and when the doors opened he stepped forward and held the door open for her and said something along the lines of "I hope I will see you around, maybe we could have some lunch?" and she said in a return flirty manner maybe. She stepped out into the sixth floor and turned down the hallway as she knew exactly where she was going (obviously she didn't).

As she went into an open floor with your normal cubical style layout she saw two people get up from their desks without locking their machines. She made her way to one of them and plugged in a USB drive that had her malicious payload on it. A few people looked at her she explained but no one said anything or attempted to stop her. In that story, she went on to make noises and make it obvious that she was not in the right spot as she wanted to get caught it was part of her tests that she wanted to know what would be needed to enact a response.

Paula's story reminded me of a time in which I had tested out physical security for a company, I had walked in, sat down in one of their meeting rooms. I then plugged my laptop into one of their network ports and started to fish around their network. I then attempted to break into the wireless network and the whole time no one was even curious why I was there. I could have sat there doing whatever I wanted. I have mentioned physical security before but I thought it needed to be brought up again as honestly what would your staff do in this situation?

Do you think they would say anything, do you think any would even bother to check if you are authorised

to be there? I don't think in most mid-large organisations that most people would even take any notice.
They would just shrug it off as another new employee
and wouldn't think twice of it in most instances. In
SMB's it is more likely that an unauthorised person
would be called out for being somewhere they are not
authorised because everyone would know each other
and would be aware in most cases that a new person is
starting.

This needs to be something that is discussed more
within security awareness training, we need everyone
to notice these things and even if they are not comfortable with confrontation but be able to advise someone
that there is a suspicious or unknown person accessing
restricted locations. That would be enough. If we can
do this it will greatly improve security and it only requires awareness. Very cheap and effective protection
to implement don't you think?

So that covers people walking into unauthorised
locations but what about if you are in a retail store or
some type of public accessible area that has PC's of
some kind in them. Could someone just walk up and
plug in a modified USB charging cable? During another
talk at the conference, I watched a demonstration from
Kevin Mitnick who used what they call a ninja cable
to execute a malicious payload to a machine and then
give him access to this via his other computer, it's very

simple and a threat vector that should be discussed with staff.

This is not a new attack vector, as previously malicious actors or Pentesters would plug in USB drives to get their code onto systems. One example of how this could occur which he discussed in his talk was sending someone a free iPhone, a prize for something they made up and replace the original charge cable with the malicious one and when they go to plug the phone in to charge. You execute the code and bam you're in. Would your staff notice a USB charge cable or think that was even a threat, no they wouldn't.

So let's change up the awareness training to ensure that it includes these threats and outline the best ways to counteract them with your staff. Anything we can do to minimise our risks the better we will all be and the more sleep we will be able to get at night.

26

WOULD YOU USE A STRANGER'S USB CHARGER CABLE?

USB charger cables for our phones, tablets, watches or whatever gadget we want to strap ourselves to these days (yes I said that correctly – strap ourselves to, not strap to ourselves – Think about it, it's probably closer to the truth). The charger cords, most of us have several for all the different devices we use regularly, sometimes they break or we lose them. It's true isn't it or you just run out of charge and you don't have one with you. What happens in this situation? Would you borrow someone else's charger cord? Of course, you would.

Would you just grab one from one of those vending machines, 2 dollar shop or corner stores? Yeah, you would because it's just a charger cable there is no risk. Most people think this and honestly, I don't blame anyone for thinking this is harmless, nothing could go wrong with this scenario.

I want to introduce you to the O.MG cable by MG. essentially the cable has been created to look exactly like a legitimate iPhone charger lead, nothing fancy, just exactly like the cable and seriously these guys have done a great job it looks perfect. Now if you look a little closer and pull back the USB plug circuitry cover you will see a giveaway, these cables originally surfaced at Defcon in August 2019 when MG showed the capability he and his team had been working on.

Look these types of cables aren't a new idea but in my opinion, these would have to be the best attempt yet, they look very real, they work very easily and are controllable via an app on the malicious actor's phone that connects from up to 300ft or almost 100 metres away via a built-in wireless adapter in the USB cable. If you check out the video that MG included in his tweet on 9th August 2019, you will see a basic demonstration of what it can do.

So let's create a bit of a scenario now so it all makes sense. It's a Friday afternoon and you are sitting in an airport lounge area waiting for your flight home to see your family, your work phone is flat and you want to just get out a couple of quick emails before your flight starts to board so that when you get home all you need to worry about is spending time with your family. You are looking around and you see a girl in her early twenties roughly sitting across from you with her laptop open working away. Sitting next to her on the chair is an iPhone charger cable, you look around a bit more and see one of those USB charge points that we all see starting to pop up in cafes and airport terminals.

You consider asking the girl if you could borrow the cable, you know that you should never plug a USB into your machine that you find or is given as part of a give-away, the risks are too high that you will get a virus or something. There is no risk with a charge cable right, you ask the girl and she smiles and says "sure go for it". She passes it to you and you make your way over to the charge point and connect up your phone. It starts to charge as expected and you wait a few moments before switching it on. Now at this time the super help-ful girl has connected to the O.MG cable and is loading malicious code to your device for some nefarious pur-pose. She will soon have access to everything on your phone, work emails, texts, internet banking passwords you name it she now owns you.

Now I am guessing some of you are sitting there thinking that's not a realistic scenario, well let's think of another one. You are at the office and your phone is flat, you sing out to one of your co-workers asking if you can borrow their phone charger cable. You grab it and plug it into your pc USB port and connect to your phone. Now the part you didn't know is that the cable was bought online in bulk, they bought 10 of them for $1 each as they keep losing their cables. Those cables were O.MG malicious cables and the staff member was targeted because of where they work. You have just given the malicious actor access to your company network via your laptop and most likely access to everything on your phone from the company car park. Whoops right.

Another scenario could be, say a doctors surgery, if you have been to one lately most have PC's on a bench as you walk in, some even have a pc for people to self-check-in, if you plugged in one of these charge cables, would anyone pay attention? Maybe. What about if you were in the doctor's room or nurse station and they turned around to get something or left the room. Plugin the cable leave it and you are in most won't think it's a threat even if they think it's unusual. They will probably just keep it and use it themselves.

All of these scenarios are very real possibilities and we all need to include these types of scenarios in our awareness training programs. Let users know of the real threat, it's about more than USB drives now. We need to inform staff and families to not use cables that they have purchased themselves from a reputable source. Look even in this scenario you could still fall, victim, if a supply chain is Brocken and a malicious actor finds a way to say "man in the middle" a new iPhone delivery, take out the charging cable, replace it with a new O.MG cable and then reshrink wrap the box. Doable right? Okay, that is probably a pretty sophisticated attack and is probably unlikely but still possible.

What about the brand new iPhone you just won, in a competition that you didn't know you entered that just arrived. Yeah, it thinks it could be done.

I don't want to be all doom and gloom here. Just do yourselves a favour and think before plugging in random phone charge cables you may regret it later. Keep one with you, have a couple and if your phone still goes flat sometimes and you don't have a charger cable just disconnect for a while it will probably help your stress levels. We all should do it now and then.

27

ZERO TRUST SYSTEMS A WAY TO SLOW THE SPREAD OF A BREACH

Cybersecurity is a really tough job to get right, we have a never-ending spread of infections or attacks and it would appear that nothing is working in the fight to stop it. Are we going about this the wrong way?

Most organisations work on a Castle and moat scenario whereas I am sure you can guess you have a large number of perimeter defences to stop attackers from getting into your castle. You have the moat around the outside of the wall to make it very difficult to reach the castle walls to attack it, you will have one funnel point of access (or two) that allows you to direct all attackers into a smaller more defensible section of your wall that you can have many defenders and protections in place to stop any attack thrown your way.

This method of defence is solid and is the main approach used by most defenders for a long time now, but this is no longer a viable option. Perimeter defences are becoming useless to any organisation that is trying to protect their employees and their company/client data. Why? It is simple, how can you protect a wall or the castle itself from attack if there is no castle? Seriously let's think about this. 15 maybe even 10 years ago companies had a business site where all staff went to work and once the day was finished they closed off their machine and went home. All the data was at that site on the one system and the walled protections around these networks were perfect.

In today's environments, there is no real castle to protect, the company's employees and assets are spread all over the globe or country or even if it is just one site your business operates from primarily your servers or primary business applications are most likely cloud-hosted. Your staff will nearly all have emails or some other business data access on their phones or tablets or whatever the latest gadget is that everyone is using at the time. So, if there is no boundary or site location how can you put up walled protections? Honestly, you can't.

We all need to look at our systems differently, we need to look at them as though they are a fluid environment with no beginning and no end. We need to think of every user and every device as untrusted because we can't just put that wall up as we used to, we need to believe that everything could be a malicious actor. Now don't panic, I know that now means you need to be in a hundred places at once but there are ways we can minimise this mess.

Zero trust is not a new idea and some great platforms can help create such a network, but I am not going to talk about those specifically as honestly, I am not a salesperson for whatever blinky light solutions they are selling. I want to talk you through some basics that will help you build this type of network that is resilient to allowing malicious actors to spread unabated through your organisation's systems.

How about you first look at segregation? Break up your systems into departments or locations, isolate them from each other virtual networks or physical network breakups and stop them from seeing each other completely. They don't need access to each other's systems so why do we allow it? We shouldn't break them up, give each area/division/site access to what platforms they need to do their jobs and nothing more (I will come back to the nothing more point in a moment). Now you have the smaller segments, put your walls up around them to protect each of them. That is a good start.

Next, let's talk about the nothing more than required statement above, each user will need to have access to a set of things to do their job, they may like or even want access to other things but they don't need them to do their jobs. So, don't give them anything more than the bare minimum access they will require to do their jobs. Close off shared drives don't allow them access to different applications that are used across the company. If they don't need it they, don't get it period. It is a simple theory really if a user is infected with a virus or falls victim to a malicious actor then if that user has restricted access it will minimise the ability for the malicious actor to gain access to company data. Yes, they still have access but the more restricted we can make it (without making it painful for users of the systems) the better we can slow the spread of a malicious actors reach. We may lose one division in an attack, but we may still be able to protect the remaining ones.

Now you have smaller pieces to protect and users only have access to what they need too. A good start don't you think? Now I am going to tell you that none of this matters really because malicious actors will still breach your networks and will still break things that you don't want them to, that's what they do and at this time I don't see that changing but that doesn't mean you should do the above. No, it means we should do it as best we can and make it so hard to move even an inch on your systems that it irritates the malicious actor so much that they just give up.

To do this we need to ensure the above has been done well, then we need to have systems that can monitor the network for any new device or software that hasn't been strictly approved and throw it back out as fast as it came in. We can do this with a good IDS/IPS with systems with that beloved AI that is touted around these days (probably just machine learning not really AI – but that is a different argument for another time). You could also use application whitelisting to stop unauthorised applications if users can't install and run applications that you didn't vet first then that will greatly reduce your threat surface.

Why don't we add in user behaviour monitoring why we are at it? Some SIEM's can help you with this, learn how they use the systems, where they would use them from and be notified if something isn't normal. It may be that missing piece that could stop a breach. If your team normally works between 6 am until around 8-9 pm on occasion but you suddenly have a user that is logging in and accessing files at 3 am then it's likely you have a problem.

Look there is many things you can do and many systems that will help you do it but the point I want to make here is simple, castle and moat protections are over they don't work. We need to always remember that every device could be malicious and design our protections that are based on the fact that the ones we want to protect our systems from are wandering around our systems as we speak. Systems in our network are no longer trusted.

Keep your fingers on the pulse, learn your systems behaviour's and respond as needed but more importantly do the groundwork so you can scrape back an ounce of control in the ever-evolving environment we are pledged to protect. Zero trust systems are not a silver bullet solution to all of our problems in cybersecurity (we have way too many to work on for that to be the case), but it can certainly help reduce our risks and allow us to respond faster to breaches (6-12 months is too long for breaches to undetected) that's a good step forward.

28

WHY DO WE NEED TO HELP USERS BE BETTER PROTECTED AT HOME?

So, you have a great security platform, well harden systems, regular updates regime and a very effective security awareness program with both classroom-style training and online training for your users. You are smashing it out of the park and making a difference in keeping your organisation better protected. That's fantastic. Are you missing a massive vulnerability though? I know what you are thinking, I have a strong security program and I have a great supply chain review process. I don't see any vulnerabilities that we are not covering.

What about your user's home and how they protect themselves when they are not in the office? Shouldn't you help your staff be smarter and better protected at home as well? I think you should consider how you could help users be better in this area. Why? Let's lay it out for a minute and think about it.

Many users now have at least one or two computers at home, they have tablets, phones, IoT smart devices and possibly even connected fridges or fish tanks. We all know that many of these devices come with threats that could cause a breach that's why on the corporate networks we isolate IoT devices on a different part of the network (well we should anyway) that reduces the impact they could have if they are accessed by a malicious actor. Do your users use a smart printer that they can directly print to from an app on their phones? Yeah, I am sure that some would but do they know that they need to change default logins and enable security features that could protect them from an attack.

What about the smart speakers or google homes or Alexa devices that many early adopters are using? What about the connected doorbell? Could that allow malicious actor access to their network? It certainly could.

This particular Ring doorbell issue has been patched but I am certain that there will be other vulnerabilities found that could be just as damaging (and a lot of users probably won't know about the issue and not patch it). Look these are just a few issues I have plucked out of my head, there is plenty more that could be a catastrophe if it was on your corporate network. It's not though you say, we are talking about user's home networks and systems. This doesn't affect our corporate systems. Actually, it does.

Many users do regular work from home, they may have protected devices that have been provided to them but they will be connecting those devices to their home Wi-Fi or LAN environment. Do you have control of this network? No, you don't. Do you even know what is lurking in the dark corners of these networks? Could a malicious actor be waiting for the perfect opportunity to strike? Are they constantly poking and prodding your corporate devices to find a way in? It is entirely possible. They probably didn't target the user's home systems to get access to your business but I can guarantee that if they see an opportunity that will be fruitful they will take the opening to attack.

They have no boundaries, no rules governing them. They are in it for the money or any potential gains they can get and they aren't biased at all, they don't care how nice a person you are or who you have to support. Do you believe me now that we should consider ways that we can help educate our users about being better protected at home as part of our awareness programs? I am not saying that you all need to take control of user home networks or support them as part of your help desk services (obviously if that's how far you want to go with it, go your hardest), what I think you need to do is possibly provide some training on possible risks that users need to consider at home, educate them about turning on security features on smart devices about the importance of having at least a good anti-virus program.

I would also suggest that you may even want to offer licences for your security platforms (AV) to your staff, it would be a minimal cost to help protect your users at home and could help prevent an incident. I know budgets are tight and that may not be a possibility but maybe worth the costs if you can handle the expense.

Think about this though, if your team can help your users be safer at home that education and security improvement will flow through to your user's corporate behaviours. That small cost that you incur to do this could provide the difference needed to prevent them from being a victim of a breach at home and also being a more secure user in the corporate systems.

So, let's go back to that first question, should you help your users be better protected in their environments? Of course, you should, I feel that all the benefits will greatly out weight the costs, offer your staff a 1 hour/30 minutes remote sessions with your help desk team each year or quarter or whatever you feel is enough to help your staff be more secure, teach them don't just do it for them. It's all about the education process, help them to be better with their own devices. Help them know where to get apps from, how they should set up devices and that they should reach out for help if they can't do it themselves, this doesn't have to be the company IT desk, teach them the value of setting things up securely and they will be happy to pay for external support to help.

The value you in investing in your staff/users speaks for itself, if not just for the security benefits what about staff morale. If your staff feel that your organisation wants to invest in protecting them in their personal lives as well as in their business lives they are more likely to be happy with their current work environments and want to be a part of that company. Staff retention is a hard thing to achieve and a simple gesture such as this could be that tipping point.

29

HOW TO RUIN A MALICIOUS ACTOR'S DAY: GET BACK TO BASICS

It's a hot summer's day, I can see the storm clouds building outside my office window. It looks like it could get real nasty outside, I hope that I don't get wet on my way to the train station for my commute home that would make the trip home pretty miserable. I am pretty excited to get home tonight, it's my daughter's birthday and we are having a bit of a get together for the extended family. Nothing to fancy, some cake a BBQ, just a good old fashioned Australian get together. She is turning five and is very excited about the whole thing it's not often she gets to see all of her younger relatives and tonight she will be the centre of attention.

I pack up my stuff and make my way to the elevator and select the down arrow to call an elevator to my floor. I stand there waiting patiently with an invisible buzz of excitement bursting to breakthrough. I see in the corner of my eye one of my security team heading down the hallway towards me with a bit of determination and focus on his approach. The elevator doors open in front of me but I pause, Dean could want me for something important. I want to get home but I won't be able to enjoy the party or the rest of the weekend if I don't ensure everything is ship-shape here. Live of a CISO I guess. I see the doors closing in front of me again and do a slight sigh from the disappointment that I was not on the elevator head out for the weekend.

I turn and greet Dean as he approaches, he nods "Boss, I just wanted to let you know that we have been experiencing some networks scans today. Looks like someone has been looking for vulnerabilities in our systems. They didn't stick around long, looks like they didn't have any success." I consider what he has said for a few moments "Make sure the team does some systems checks for anything suspicious and have the night weekend security team monitor for any further probes of the systems." Dean nods "have a great weekend Boss, see you on Monday" he turns and heads back down the hallway. I lean forward and press the down button for the elevator again and the doors open almost immediately. I enter and press the ground floor.

This scenario could have ended so much worse, Dean (the security guy) could have come to inform him of a breach or that a ransomware virus was spreading through the network like wildfire and consuming everything. Nothing they have done or can do is stopping it. Yes, he could have done that but in this instance, the systems that were being bot scanned (the usual method) did not find any methods of access. There were no missing systems patches, all unnecessary ports are closed and systems hardening has taken place.

The computer systems have been maintained well and all of the basics have been done well, you know all of the mundane things that many don't spend much time on because that's not exciting? We all know what they are as no one likes doing them but if the scenario/situation was the breach has occurred and all hell is breaking loose our CISO (we will call him Jim) would have not been able to leave for his daughter's birthday celebration, major interruptions and brand damage would have likely resulted.

Most breaches occur due to a known vulnerability that has already been patched, poor password security or just sloppy systems hardening. Most breaches are not some exciting Hollywood style big sophisticated hack. Okay, some in the wild may be but it is rare and nothing as exciting or as dramatic as it is made out to be in Hollywood. If they showed it how it happens you wouldn't be able to give tickets away to a movie about it, it's just not that exciting on most occasions.

So let's do a quick recap, most breaches are caused because the basics aren't done correctly. So why do more businesses not do better at this? I feel it is our human nature that Is the problem here, we don't like to do boring mundane tasks, we like blinky lights, we like new toys, I think it's that simple most times. How about we change that, share the load for the mundane tasks, put in a patch management platform and use it, check that it is doing what it is supposed to. Do systems hardening and practise safe password practices. If we can achieve these basics you can ruin a malicious actors day, week or even year if we can all come together and stop making it so easy for them.

Please roll up your sleeves and get it done. We don't need to keep saying this over and over its not fun for me and I am sure that you are getting sick of being told that you need to do the basics better. So be like Jim, cover your basics and enjoy your time with your family and friends not cleaning up a mess that could have been avoided in the first place.

30

WHO WANTS TO START A RANSOMWARE CRIME GROUP WITH ME?

I serious who wants to start a ransomware crime group with me? I have been looking over the earning statistics for 2018-2019 while updating my user awareness training deck and ransomware alone cost businesses more than 8 billion dollars each year. Wow, that's a lot of money. Are we on the wrong side here, should we go out and set up a nice ransomware business, we could even run it like a legitimate business in a country that has no extradition and make a real killing.

I have read many articles and blog posts over the last year saying that ransomware is a dying threat but I don't think it is, every month there are more and more companies admitting that they had been affected by a ransomware attack and them not knowing how long it will be until systems would be back operational. Many pay tens of thousands to get the data back and this is not a factor that is going away.

I don't think it would take much, we could buy a good quality ransomware platform on the dark web, set a spam centre/call centre scenario and only pay on results so we just take a big cut of the profits without actually doing the work. It would only take a few months and we should be rolling in the money. I am not sure I like most of the locations we would have to move to, I am a big fan of Australia personally but it could be a big windfall.

We could always buy big mansions with fast cars to go with them, which would make the moving country a bit more bearable don't you think. There is one downside though, we would have to become a slimy, horrible criminal who has no regard for anyone at all except the growing purse strings. We would have to ignore the fact that we would be destroying the businesses of hard-working people who work 80 hour weeks to keep their businesses afloat to just wipe them out in one fell swoop never to recover.

A small business that can't find out who owes them money or has any information on most customers will not survive very long. Children would go hungry with no money or ability to earn any. Families would break apart from the effects of stress and suicides real possibilities. Could we live with that? These people do, many would even lose an ounce of sleep, living it up in their mega-mansions, showing off the money and power they wield.

I just don't think it would be something that I could do, I guess like most of you I am destined to work a day job (unless one of my books gets made into a movie or something crazy like that) as I will not do that to people. I believe myself to be one of the good guys fighting to rid ourselves of a world in which people that find no remorse in the affliction of such devastation to people's lives. We need to find a way to help people be better prepared for these types of attacks, teach them how to recover and prevent total system losses.

I know if we all come together and focus on how to make a difference we will see a decline in the number of incidents but I know deep down that this is a fight that will not be won easily. I know I was joking around saying who wants to start a crime gang with me but this is a very serious problem that has hurt a lot of people and not just financially.

We need to cut out the lingo, come up with a simple way to get the message across and help people to not recover (yes this still may be necessary) but to prepare for the inevitable incident and make sure that they have those basics covered and well set in so that a ransomware breach is just an annoyance and is something that in a day they could recover from with a bit of work.

Just to leave you with a final thought if you do see me in a Ferrari or mustang (I would love an old school Shelby mustang personally) it's not because I have turned to the dark side that's not going to happen but it will mean my upcoming hacker fantasy book "Foresight" is gold and it has been picked up by some Hollywood movie studio (that would be awesome) but know this, even if that did happen I would still fight this fight to have a safer cyberspace, a better prepared and educated community.

That's my dream to make a difference. So let's put our heads down and get figure out how to stop this pain in all of our side's ransomware problem. It just really is painful.

31

SHOULD I CALL MYSELF A CYBERSECURITY EXPERT?

The word expert is in my opinion thrown around too much, we have expert cybersecurity professionals, Microsoft word or excel experts, email experts and that's just thinking about the IT or security field. There are so many different experts in whatever people decide to call themselves experts in and honestly these people don't always ooze professionalism or skill in the chosen areas, so why do people do it? Simple, I think its egos and money that make people feel they need to make themselves stand out from the crowd in their made-up niche area of expertise.

Look I am not saying I don't know quite a bit about information systems security/cybersecurity and even IT in general as I have been involved in IT since the early 2000s and cyber for the past five years. I have two master's degrees one in IT management with a digital forensics major and one in Information Systems Security. I have a bunch of industry certs to boot but does that make me a cybersecurity expert. MMMMM I don't know if it does, yes it makes me knowledgeable and a clear member of this awesome industry of ours, but I don't know everything in security or will I ever say that I do.

I don't think anyone does know everything about everything. I have some areas I probably know more than others and have great general coverage across many areas, but does that give me the right to wear an expert badge.

The dictionary definition for an expert is interesting though: a person who is very knowledgeable about or skilful in a particular area. If we are to go with that clear-cut definition, I would fit that criteria and I guess most people who classify themselves do meet that statement, they would be knowledgeable of their area in which they specialise.

So, my resistance to being called an expert is not in its definition but more in my reservations about the idea of being called an expert. It comes down to probably imposter syndrome or similar in which we lack the self-belief or confidence in ourselves to allow us to accept that we are in fact experts in our fields.

Do you like to be called an expert? I get a weird resistance to the name when I am called that by a peer or work colleague, but I really shouldn't. I should own my abilities, Yes, I could always learn more but we all can. We will never know everything and anyone who tries to say they do are lying or have an arrogance issue.

I don't want to ruffle anyone's feathers with this, this is just my opinion on my thought processes around being depicted as an expert. I prefer to be identified as a security engineer, ethical hacker, journalist or even author is fine. Just not an expert. I have probably opened a can of worms with this chapter as people will (just to stir me) probably start to introduce me as a cybersecurity expert. I guess I should probably get used to it.

So if you are honestly knowledgeable in your area, don't shy away from the expert title as I would normally do. Claim the title and strive to earn the name, keep learning, practise your trade and just be humble.

As always tell me what you think and share your thoughts/feelings around this, if you too resist the expert tag? Let's help each other break out of our shells and shine.

32

EAR PODS AND WIRELESS HEADPHONES: COULD I EAVESDROP ON YOUR CONVERSATION?

I have considered this question a few times, I see them everywhere nowadays. Being a security engineer and penetration tester (ethical Hacker) I admittedly run through scenarios like this all the time it is just how my brain works, could I get access to that, is that something I could manipulate, did they put security on that device. I have done some travelling on public transport and you see things that just really puzzle me sometimes, people do things in plain sight and in reach of random strangers without any consideration of the risk they could be putting themselves in.

I am going to take a closer look at the Bluetooth and wireless headsets question that is the title of this chapter and I will also look a little closer at some of those risks that you may not have considered that I feel you really should. I have touched on them in some of the other chapters but there are some really important points we need to consider so will run over them again as needed.

Let's run through a scenario as we all know I like to visualise problems and scenarios to help everyone understand a point of view or situation. It's a normal workday and you are on your train to the office. You have your new whiz-bang earbuds in your ears and you are listening to a podcast from your phone. It's something you like to do all the time, listen to a podcast about life or something funny depending on your mood. Today you are in a self-reflection type mood so you have opted for a bit of self-help style podcast. You are about 20 minutes into your 50-minute train ride to the city when your phone rings, it is Dave the finance officer at work, and you answer it with the usual pleasantries between each other. Dave asks you to confirm the bank account details of a new hire (you are the HR officer at your organisation) as today is payday and he is missing a number from the account number he believes.

You are on the train and you decide that you don't want to read it out on the train so you tell Dave that you are travelling to work and will get it for him in 30 minutes once you reach the office. He tells you that he needs it now so he can process the pay run otherwise he will have 1K people yelling at him because they don't have their pays on time. You consider it and advise him to read it out to you and will confirm if it is real or if numbers are missing. Dave reads it out and there is a number missing off the end which you advise him what it is.

Just to be certain that all details are correct, Dave reads you out all of the staff member's info like date of birth, full name, address, contact numbers and even the emergency contact info. You end the conversation with the same usual pleasantries as you did with the start. You (we will call you Jane for fun) just go back to your podcast none the wiser that a malicious actor had been listening to the whole conversation you were having, they had intercepted your Bluetooth signal and have been capturing your unencrypted signal. This was possible because Jane had bought knock off earbuds and the manufacturer cut costs by not implementing any security controls on the device. This is not uncommon with many low-cost devices that are made to be cheap and easy to configure.

Many of the better quality devices like the ear pods will use a 128bit encryption of the communication via Bluetooth and will also have some basic security authentication handshakes which are in most cases more than enough. But if you have some of those devices that have cut some corners to save a few cents or even dollars on a device. I get it everyone is in business to make some money but we all need to do it right. if someone had been listening to Jane's call and could listen to both ends of that conversation that is a lot of information about the new employee. The person listening in could use that info to break into that person's accounts online, maybe even enough to get into their banks and drain them of their money. This is just one scenario though.

Think about what conversations happen all the time in front of you, I hear people have full-on conversations that are confidential and they seem to be oblivious to the fact that they shouldn't be doing it in a public place.

I bet some of you are reading this thinking that you don't buy cheap headsets or earbuds this doesn't affect you, it still does. Encryption or good security practices with devices is just the first battle. Bluetooth is a massive standard that consists of many different variants with components written and reused by different people. Many of these millions of pages of code could lay a vulnerability that is just waiting to be exploited that could easily allow such an attack with ease. They could capture all conversations in an area around their device to later go back through and divulge any useful info.

They could use the vulnerabilities to install a sort of bug on your device so that it can listen to you or anyone around you via the microphone. That type of attack is nothing new, you hear about it happening all the time with computers why not mobile devices. Personally, it makes more sense to do this on a mobile, these days people are attached to them as their life depends on it. If they have them everywhere they go just imagine the information a malicious actor could capture from a target. Why break in when they can tell you everything that you need to know.

Stealing company secrets or spying on other countries with this method is something that you can't tell me hasn't occurred before now. Look I don't think we should all be paranoid about everyone listening in on our conversations around us using our mobiles although apple, google and amazon (I am sure there is more of them) all been listening to recordings made by the smart assistants supposedly for research to make them work better but chatter around indicates that they have been listening to much more than just when we talk to the smart assistant which is pretty unethical so make sure your turn off the settings that permit them to do this. It's hidden down deep in most smart devices but it is there if you can find it.

I just believe that we need to be more cautious about how and when we use these types of devices, I still feel that there isn't anything wrong with a cable. Back to my point, look at the security component of your devices when you purchase them, don't have conversations that are private in a public place like public transport or cafes or even an intersection while you are waiting to cross. Honestly, you never know who could be listening or who could be watching what you are doing while you're glued to that screen or have an oblivious conversation telling the whole world your personal information.

Think about your security, it will make you much more secure...

33

LARGE EVENT CYBERSECURITY THREATS: WHAT YOU SHOULD CONSIDER BEFORE GOING TO YOUR NEXT

Have you been to a concert, music festival, car racing, conference or any other large scale public event? If the answer is yes to any one or more of these I would like to put out some scenarios for you now and help you to understand some risks that you should prepare for before you head off to your next one.

Let's pick one of the events mentioned above, let's go with a conference. At a conference (I regularly go to cybersecurity conferences which probably escalates the threat level – a massive venue full of budding hackers – sounds like a perfect place to get hacked), we will go with something a little less risqué and go with an accounting conference. Now I will be quite honest here, I don't know how wild accounting conferences could get I have never been to one but I am going to bet that on the cyber threat risk level it is going to be lower on the scale as far as conferences go compared to a cyber conference. I think most of us could agree with that, however, if you are an accountant and want to tell me how crazy it gets please do.

For now, though we will go with the accounting conference. Let's say James is an accountant he is heading to Melbourne from Brisbane for the 3-day event. He is going to take his work laptop and mobile with him so that he can still do work around the conference events over a few days. He makes his way from the airport to his motel that is above the conference facilities and has been given access to the free Wi-Fi during his stay. He is given a swipe card that will give him access to his room and all other allowed areas of the motel including gym, pool, guest-only dining facilities and the underground car park if he parked a car downstairs.

He flew down the night before the conference so he was nice and refreshed for the events over the first day.

He goes to his room, orders some room service and connects up his work laptop to the motels free Wi-Fi. It's a little slow but works okay. He connects up to the office remote desktop server and does some work before his dinner arrives. After around 30 minutes he hears a knock at the door, his room service has arrived. He leaves the laptop connected and eats his dinner and decide to watch some TV to let his dinner settle down a little. While he was eating his dinner though James didn't know that his laptop was connected to by someone else on the same free Wi-Fi connection.

They had already captured all of his communications to the remote server after he connected to the remote server as the Wi-Fi that James connected to wasn't the real motel Wi-Fi connection it was a duplicated id boosted by a pineapple device that allowed the malicious actor to pretend and pass on any communication between James and the internet. After the details had been captured it was easy for the malicious actor to connect to James laptop it uses the same account as he logged into on the remote server. The malicious actor would now install a remote access method so they could gain access at a later time (maybe so they can spread access onto the corporate network once you go back to the office).

Okay so that was a little Hollywood but very possible given the right circumstances and James would not have been aware of the incident. In this first scenario,

James hadn't even gotten to the actual conference before he had been breached. There are a couple of quick lessons from this scenario, don't use free Wi-Fi (ever), when travelling make sure that you use a VPN to protect your communications from eavesdropping. Also make sure that your organisation adopts multi-factor authentication, in today's environment it's a must.

Let's pretend for a moment that James wasn't breached on the first night he arrived and made it to the conference first day unshaved. He is walking down into the conference area and a nice young lady is handing out iPhone and android phone chargers with what looks to be an accounting software logo on them, it's packaged well and looks good quality. He takes the cable, it never hurts to have a spare charger cable now does it.

James goes through the morning with no issues, he quite enjoyed the presentations and makes his way out to the café near the front of the motel, and he opens the laptop and decides to do some work before he comes back to the next session. He looks over at his phone and it is almost flat, there isn't much signal in the conference area so it is burning through the battery trying to get a signal all the time. He remembers that he got that charger cable this morning, he takes it out and plugs it into his computer and then connects up his mobile. While he was connecting it up to his mobile he didn't see a command box pop up and then drop always just as fast. A malicious actor now has remote access to his machine.

The USB cable was a way of getting access to the machine, it had a hidden chip that installed the desired applications on connection. All without the user knowing anything about it. Just imagine how many of these cables the girl could have given out to people walking into the conference. This could be done at any event and be very successful as people wouldn't see a charger cable as a threat but you need to. Don't accept cables or random USB drives. It's not worth the risk.

Let's think about financial risks while at an event like this, most people have RFID credit cards that allow users to pay wave, do you know that a criminal could walk around with a handheld device that could take payment for say $99 just by walking past you and scanning for payment cards. On EFTPOS machines you would hear a beep but if that payment terminal is inside a backpack and all the criminal had to do was hold an antenna in their hand, small not very easy to see would you be able to tell if they had taken payment from you in a noisy conference or music festival, no you wouldn't until you noticed the payment on your account a few days later.

This type of theft can occur in any public place or public transport. Crowded areas in which you won't notice someone coming up close behind you. Get yourself some RFID shielding sleaves before you go to an event they are only about $1 each and could save you from the hassle of trying to get your money back.

Mobile phones are also something that you should secure before going to an event, turn off the Bluetooth to ensure that you don't get unwilling paired with a malicious device and it may even be a benefit to turn off the Wi-Fi so that you can't auto-connect to any rogue access points.

Most of what I am saying here is not new, not even close but if you think about the risks before attending any large public events and do the right preparations you could save yourself a lot of drama or financial pain from falling victim to a malicious actor. Oh and please stop using free Wi-Fi, data is quite cheap nowadays, use the hotspot on your mobile or purchase a cheap data pack for you're to use if travelling overseas or just going to an event that requires you to still stay connected.

So think about your risks and be prepared, it's that simple.

34

BIO-TECH SECURITY: MY THOUGHTS ON FUTURE RISKS

Many of us have grown up with movies and books depicting an amazing future in which we are all zipping around in flying cars, have robots that help keep our homes clean, instant meals at a press of a couple of buttons and in some cases a world in which man and machine become one. Some favourites I can think of is Robocop, I-robot, Ex Machina, Elysium, Terminator, real steel, chappie and so many more there is too many to list. Some of these are tales in which humans are targets of a robot takeover plot which I guess is a possibility also but some like Robocop or Elysium are when we use the machines to make ourselves better, stronger, faster.

I know that is all Hollywood drama and reality will be very different but when it finally does become a reality that humans can become as much machine as human (although it is 2020 which is when many of the old sci-fi movies indicated this would be reality), it will dramatically change our lives, some for the good and some for the bad.

Amputees who can just receive a new arm or leg, will probably increase their abilities instead of impeding them. Individuals with spinal issues could be given full functionality back to their limbs or have them aided by artificial limbs. I have seen many porotype exoskeletons that could be great for this type of use case. These are primarily pushed by the military at the moment due to the ability to almost make a super-soldier. Another vision born out of science fiction. Strangely, the number of inventions that have been made because it was put in a movie or book is huge. Someone somewhere was sitting there going "I can make that a reality, I can do that". That's all it takes is an idea and someone with the determination to make it happen.

I think the possibilities are extraordinary, along with flying cars (can't wait to have a go of one of these) and I believe it will happen to a large extent in the next 20 years but although I am excited for this change to occur it brings with it many security concerns. Some that I think we as humans need to consider before reality sets in and it is too late. Once machines are part of us, being hacked will become a real threat. Think about it IoT and computer systems are under constant attack, the defenders are not winning this battle.

So the idea of making my body part robot scares me a little, how will we ensure that the robot/computer side of us is not breached. Imagine being ransomed to gain control back of our bodies, you would have no choice but to pay or they could wreak serious havoc on our daily lives. If they could gain full control and you have multiple artificial limbs they could make you go wherever they desire. A kidnapping would be so much easier if they could make you go to them don't you think?

These robotic body parts will need to have strong security in mind from the very first concepts, this can't be systems in which security is an afterthought or just a bolt-on because people complained later that it didn't exist. That will not provide the level of security that is needed to ensure that human lives are protected as they need to be.

An update program will need to be ensured and handled by a trusted party, it will need to have strong controls so that third or fourth party apps/programs are not able to be installed. This is a must or this future will start to head down the Hollywood path, with armies of everyday people being turned into slaves of an unseen aggressor (this could be an AI aggressor) a real possibility if you believe some AI conspiracy experts (loosely stated the word expert). They believe that it is only a matter of time that true AI will turn on humans and indicate that we should not continue the push for true artificial intelligence, true consciousness.

I believe that there is certainly risk and we as humans will need to consider the risks, not just dive in headfirst with blind ambitions but I feel that this AI take over will not likely occur. I think the biggest threat is humans behind an attack against any Biotech or robot systems.

So with that threat in mind, let's build these systems right in the first place, build in protections that will prevent the third party installs, only allow updates from an authorised point and make sure that constant updates are delivered and control of the artificial limbs remains in the control of the owner (person it is attached too).

If we can do this it will improve the outlook for a biotech heavy world where the line between machines and humans is becoming more and more blurred. I think it is a bright future we just need to put in the effort to ensure that it remains on that path, one on blissful integration for the benefit of mankind, not its downfall.

35

RFID SHIELDING: IS IT WORTH IT?

I get asked this question a lot by friends and family as you would probably have guessed I am the go-to IT/ security guy in my group. Most groups have someone who is their go-to tech person but luckily for my group, I do this stuff for a living. So it doesn't surprise me when I get these types of questions. However, when I was asked about RFID (Radio Frequency Identification) shielding by one of my friends and if they should invest in some protection, it became clear that many people may want to know the answer to this question. So here is a short chapter giving you my thoughts on this and my personal opinion of whether or not this is a real-world risk.

Now I want to be clear here, RFID attacks are possible and I do own some shielding sleeves for some of my cards I carry in my wallet that I use when I go to security events or conferences. Honestly, I am a hacker and I know how my brain works, I would never do a live attack on anyone like this but that doesn't mean the hacker side of me hasn't thought about, run the scenario through my mind as I walk into a room. So why would I be stupid enough to step into a space with hackers, wannabe hackers (these, in my opinion, are the most dangerous as they are quite skilled yet but want to prove themselves) and security engineers/ professionals?

It's like waving a red flag at a room full of bulls and then wondering why you are suddenly being chased down a hallway by a bunch of charging scary bulls. You get where I am going with this right? Yes in certain situations RFID shielding is a brilliant piece of $1 protection. Large events or security events I have that locked in.

Now for the everyday person who is just going to work, travelling on public transport, shopping the chances of actually having one of these attacks conducted on you are very small, I would think at this time you would have more chance of winning the lotto than you would be to fall victim of an RFID scam/attack. So in my opinion, don't go out and waste your money on RFID shielded bags or jackets or clothes of any kind, it is not worth the money. Hey if you are like me and attend some public events buy a couple of the RFID sleeves for a couple of dollars each to put your cards (credit cards, motel swipe cards, work access cards) just to be on the safe side so you aren't sitting there looking around at your industry colleagues or music fans thinking who is stealing your data?

Not to say that as the technology develops and the ability to scam more money from unsuspecting victims will not change my opinion of this need in the future but it is clear for now that this attack vector is more just a bit of fun for researchers or the likes who want to play around and just see what is possible.

The only exception to this would be if you work for a government or high-security organisation in which you could be a direct target of an attack, then my recommendation to you is to buy up big, get all the RFID shielding protection you can fit on you, install it in your home, your car, install it everywhere you can as if you are a target, malicious actors will find a way to get what they need, put the protections in place before it becomes a real threat. Okay so maybe I just went down the tinfoil hat kind of conspiracy nut rabbit hole but in all seriousness, if you are at a real risk get yourself a self a box of shielding sleeves and that will likely be enough to keep you covered as well. Ignore the RFID clothing it is just a gimmick that is a waste of money.

36

WHY SECURITY IS CREEPING OUT OF THE SHADOWS

10 years or so ago, Cybersecurity was something that was talked about in hidden rooms, secret meetings and was almost like some dark magic practice that most didn't understand or feared. There were some blips of populous fame for some would-be hackers who made a name for themselves in this almost medieval time like Kevin Mitnick but most remained hidden figures with little or no public profile.

Being in the public eye was not something that was an accepted practice, it was frowned upon in fact. Over the last few years that has started to change though especially in Australia, we in the security industry are no longer the forgotten department hiding out in the basement of businesses, we are on the board, we are at management meetings and we are constantly in front of users and the general public.

So why have we come out of the dark, in full public view and a position of influence? I think that the answer is simple, it was necessary. I think it is as simple as that, breaches were happening for large organisations and people were freaking out about their data, who had access to it and what were companies doing to ensure that information was safe. The more that we (the general public) blur our lives with the on-line and technology world the more breaches and incidents occur. The more that malicious actors would set targets on unsuspecting victims.

People live a majority of their lives in the digital space, I myself have a big online presence. We nearly all have social media profiles, blogs and even our medical data is online. Cybersecurity professionals are the defenders of this world and no one would feel safe if they couldn't see the army of people who are defending them each day.

Even ASD (Australian Signals Directorate) a once mysterious division of the army that was tasked with secretly defending our country and attacking our enemies who were mostly only talked about with hush tones. These people were the secret cyber spies and cyber assassins of Australia who had operated in absolute secrecy suddenly had been thrown out into the public view.

I think it was a great move and I would have loved to have been in the room between the government goonies and the ASD goonies when they were trying to explain why they were suddenly being thrown out into the light. ASD's move into the limelight and the formation of the JCSC (Joint Cyber Security Centres) in all major cities as well as absorbing the ACSC (Australia's public-facing cyber department/entity) cemented to the world and Australia the shift out of the shadows. It was the government almost flipping the bird to the malicious actors and other state-sponsored cyber groups. "we are here, we are not hiding and we are here to stay" an interesting move compared to how things used to be done.

Let's look at this a bit closer, cybersecurity is on the news almost every night, we need people out there telling people we are in this fight, we are here to help you fight this fight and we all need to stand together.

We need to be on the boards advising ways to keep companies out of the news, we need to be at every business development meeting ensuring that security is thought about on every new platform, ensure that it is implemented and checked at every stage. That is the only way we will make a difference.

We need to bring more people into the battle, recruits are needed at a constant rate and we need to have strong and wide diversity. We need people from all backgrounds who can look at this issue from an angle in which the traditional cyber force wouldn't have looked at it before. Who considers the same problem but can offer alternatives to current approaches, that's how we will move forward.

We need to be approachable by users, they won't approach the dark hooded shadowy figures that have been the mythical creatures depicted in both media and Hollywood. We need people to be comfortable with asking questions, learning about how to be safer online, helping them be better educated about all things cyber risk. So let us continue to pull ourselves out of the shadows, let's build our army of budding cyber people and let's show the world that we can take back control of the internet once and for all before it becomes too late and we lose all hope of ever getting the upper hand on the cybercriminals or shady governments who wish to push their control.

I may have got a little off track in this article but I am sure that you all understand what I was trying to get at here. Throw off the hoodie stereotype and let's stand together as we always should have...

37

EVERYONE IS
A TARGET

Cybersecurity still seems to be at the bottom of people's lists. Some feel as though it isn't necessary as they don't see themselves as targets. In their opinion, they have nothing of value, why would they be targeted? Why do they need to dish out for antivirus software? As Security professionals, we need to find a way to communicate the following message.

Everyone is a target. Yes, even you.

Being a target doesn't necessarily mean you have done something or have enemies. Most scanning is done by bots and then highlighted to malicious actors. If they happen to stumble upon your network or various devices and you are an "easy target", then they will breach your systems, steal your data and then if the opportunity is available; encrypt your company's data and hold for ransom.

In what seems like the blink of an eye, an entire business could crumble. Obviously, if you have the provisions in place to protect you from this then you may be safe (for now at least). Admittedly some breaches and attacks are planned carefully and meticulously for weeks, months or even years. However long it takes. They don't know what you do or even care, they just want to know how much money they can make by stealing your info or ransom from you or your company to get your data back.

Trust me on this, a lot of people pay, too many and that is why ransomware is such a profitable and popular business. People pay, they unlock your data and then in 3-6 months the cycle probably restarts. Many companies don't learn from their mistakes and vulnerabilities and don't fix the issues that caused the breach in the first instance. This is something that we need to work on, we need to help small businesses, the majority of whom will say they are not at risk to this, we need to find a way as an industry to help better educate this group. Help them at least get the basics right, don't you think?

What can we do as an industry, Nay community to help to resolve this problem? Just brainstorming here, let's host events targeted at SMBs or companies that are lacking in security staff and or skills. Presentations about the very basics of security. Run through standard policies and practices. Many of you may feel that these may be pointless, people are in the wrong profession if they are unaware of the above. False, some people may need a refresher. Attacks are constantly evolving and education in security cannot cover it all. Some small companies may not be able to hire security trained staff members. If we can help educate the general public per se, this would be of tremendous value, at least in my opinion.

We could also offer recorded sessions for businesses who register or who are unable to attend the events. Run free webinars that can be watched at alternative times. Let's invest our time in helping everyone be better protected, be better educated on risks and let's start to turn that tide of breaches around. Maybe we could put the cybercriminals out of business why we are at it or at least slow the flow of money to their coffers.

Back to my main point of this article, everyone is a target no matter your business size, yes you may not be as valuable as some of the bigger fish (targets) but money is money to a malicious actor they don't care who or what you have or even what you will lose from the target that isn't something that affects them, it's just about the money it is that simple. Do yourself a favour and talk to someone about what you need to get right, to have the basics covered. You don't need to have a big budget to make a change for the better, just put it as a priority and make time to ensure you are prepared because as many of us in the industry say "it's not if but when an incident/breach will happen".

Let's do better, be better and give all the criminals less of our hard-earned money.

38

PASSWORDS: GETTING THE BALANCE RIGHT

We have been forcing password complexities on our users for years. Whilst incredibly secure these types of passwords can be woeful. Users can't remember a good one so they make them as simple as possible or simply modify one digit on a continuing cycle and use that same one for everything, which is not secure.

We, the security professionals have caused this conundrum. We have told users for years that they need to have symbols, numbers, uppercase and lowercase, substitutions and the need to reset them every 30-90 days as well as have different ones for everything. Average users will not do this or at least will not do it very well. Doesn't that mean that we the professionals need to give advice that makes being secure an achievable possibility for users? It kind of makes sense that this responsibility falls on us.

Take a look at the lists online of the most common passwords; one I found here has a list of the top 25 for 2019. These are nearly all easy keyboard or similar passwords. Users appear to be taking the path of least resistance. I can't believe that QWERTY and 123456 are still on the list or even password. We simply need to do better.

How about we do something simple and constructive for our users that will help them help us have a more secure system. I propose something a bit strange which is going against a general security standpoint. Just breathe for a moment and prepare yourself it may be a shock.

Turn off the password complexity requirements for your users.

That's right, turn it off and make it so that users don't need to have uppercase or lowercase or any symbols of any kind. Now before you, all call me crazy and jump up and down about how crazy this idea is I want you to still keep the minimum password length. Make sure that all passwords are at a minimum of 20 characters long, make the max 30 or 40 characters. Active directory will support up to 256 characters but I think typing something of that length would be crazy. However, as we are empowering our users to control their security fates let's allow them to decide what is too long for them.

You get what I am doing by now I am sure? I am suggesting that we allow users to use passphrases for their corporate accounts. Tell them to pick four or five random words that make sense to them, spaces or no spaces it doesn't matter the choice will be theirs. If they then choose to, they can go crazy with substitutions or symbols or numbers but the choice will be theirs. We just need the length and they can mix it up a little if they like.

I know that passphrases are still not perfect and that day will come when we can get rid of passwords all together but at this time I think that a passphrase is our best choice. I also am aware that I am recommending that you go against Microsoft best practice recommendations that you have both minimum length and complexity requirements for all users. I feel that this recommendation is out of date however and we need to continue to work towards the best possible solutions for our users and our organisations as a whole.

Before you all go out and turn off password complexity requirements and jack up the minimum password lengths talk to the business about it, explain why you want to change it, get management buy-in. If you get that, talk to your users let them all know what is happening. Don't just send an email out to staff, set up a meeting, walk around staff if you can, ensure that they all know why you want to do this and then you will get user buy-in. Then they will create good passphrases and they will do it for the better good, they just need to understand. Trust me your users will want to help, they want to do the right thing. Especially if it will even make their lives easier by doing it.

Okay so that sounds like a perfect result which will probably not happen in reality but if you do it right it won't matter if there are a few bumps in the road the results will speak for themselves. More secure users, less password reuse and fingers crossed reduced risk of account password breaches.

Get out there and do the right thing for your users and change your password policies. If you disagree let me know, do you have a better alternative? I want to know your thoughts and suggestions to try and make us all more secure.

39

BE AN ACTIVE MEMBER OF THE SECURITY INDUSTRY

I have a bit of a personal belief in the benefits for us all if we can all be active members not just bystanders to what is happening around us. Let me explain this a little and tell you why I think this is important for not only us personally or our organisations but the security industry as a whole and possibly even the community as a whole.

The old way of doing things in cybersecurity was to stand in the shadows, keep hidden and you didn't share intel or incident details or anything at all especially with businesses similar to your own. They are competition, right?

Essentially yes you may be competition in the specific industry verticals you operate in but this is about more than just your products sales this is about being safe as an industry from the malicious actors out there in the dark lurking, waiting ready to jump out at a moment's notice to crush your entire system with one digital swipe of their hands.

Honestly, as I wrote that paragraph I could almost imagine a Tron like scene with this big digital arm swinging around to knock over some digital representation of your office building. Gone, smashed, wiped from the memory for good with no chance of recovery.

I live in Brisbane Australia and we have a great infosec community, one in which we regularly come together, discuss the everyday stuff. It works well for everyone involved and we start to get to know each other. In time you can build a network of people who are happy to take a call and be your bouncing board when you are in a situation in which it (the S*!t) has hit the fan so to speak. It's that sense of community, that feeling that you are not alone in this fight and the comfort if you need the help you have someone to call. That alone is worth the involvement but there is much more to this than just have that "Call a friend" option.

I have been thinking about this a lot lately, I want to make a difference in the industry and a lot of people that I talk to all over the world have that same drive I do, they want to roll their sleeves up and dig deep to make a change for the positive.

While writing this book I participated in several interviews and webinar sessions to help spread my knowledge and push my involvement to a new level. I did a podcast interview with **Garrett O'Hara** from Mimecast who is the host of the "**Get Cyber Resilient show**" we talked about my experiences, how I got into cybersecurity and even my books (This one "**A Hacker, I Am – Vol 2**", "**A Hacker, I Am**" and my upcoming hacker fantasy novel "**foresight**"). The conversation happened early on Friday morning just as I was absorbing my morning coffee dose, the conversation was great, Garret is a very engaging guy which made it easier.

We had an engaging discussion about all things cybersecurity (which encourages my introverted self to suddenly become this wild extrovert. Who knew right?). I love how easy it is to have these types of interactions though, one hour out of my day having an interesting conversation to maybe help others have an easier road into cybersecurity or get a little more insight as to any threats, patterns that I have seen on a day to day basis.

Charles Sturt University and their partner company IT Masters invited me to participate in a webinar - **How to develop your IT career into Cyber Security with IT Masters graduate Craig Ford**. It was essentially an interview of me about my experience completing my two master's degrees with CSU and how I made the transition from an IT career over to cybersecurity. The session run by Guy and Chantelle from IT Masters/CSU was great, attendees asked a lot of questions and I had people reaching out to me, thanking me for sharing my experiences with them and how much it helped them to hear about someone else's journey. How it felt for me, what worked and what didn't. Even failures, honestly we can share failures, we don't need to be perfect we are all human after all.

I have talked about this in a previous chapter in book one (A Hacker, I Am) when I discussed entering the lion's den, I am a member of the PISN group in Brisbane that was created by a guy called Andrew who is the CISO for Superloop. He is a great guy and as I do wants to help make a difference. He has this group in which he invites around a dozen or more competitors into the same room, we eat pizza, drink beer and share honestly about how we think things should be done.

We obviously don't always agree with what everyone says but we are polite and share reasonably openly. As you would gather we don't share any company-specific data or anything like that but we help each other figure out what works well and what doesn't. It is a great group and I love that Andrew continues to drive this initiative forward.

I want more of us to be like Andrew, Garret, Guy and Chantelle are in this chapter helping to bring people together, help share the knowledge. Why can't we all be more like this and start to create an open and collaborative industry? I was asked by Guy in the webinar if I thought this type of collaboration could occur or where we all just blowing hot air (I don't think he used those words but that was the gist of the question) and you know what I think we can do it.

Let's drop off this ego problem that we all have seen in this industry and let's be active members in the industry. Let's share our knowledge and make it easier for the next generation of security folk to learn from our mistakes that way they will be one step closer to achieving the win that we seem to be just shy of reaching against these persistent malicious actors.

Go do a google search for the podcast by Garret – **Get cyber resilient show** and for the recording of the webinar for **IT Masters - How to develop your IT career into Cyber Security with IT Masters graduate Craig Ford**. Something that I have shared could help you or it may encourage you to be more open and sharing. Just think of it as an interactive multimedia addition to the book.

Honestly, though go out there and be a real member of the community, whether that is Cybersecurity or not, share your knowledge to help others make an entry into your industry and don't shy away from mistakes, share them help others and yourself learn from them. Trust me you will be better off for it and so will your industry no matter what it is.

40

THINGS YOU CAN DO AT HOME TO IMPROVE YOUR SECURITY

This chapter is going to be focussed very much on the home user scenario, I want to give some advice on what everyone can do to keep themselves a little better protected online and prevent some security breaches that could have been prevented with a little bit of work. Now some of these things I am going to discuss may be out of your abilities and that is okay (don't stress if you don't know how to do it or it is above your skill level) ask for help, get a mobile technician out to your place and invest in making sure you are a little safer moving forward.

Task one to tackle is making sure that your computers are all up to date and ensure that you know how to do this moving forward. Yes, sometimes updates could cause minor issues but the security benefits are much higher than the costs if something were to go wrong. Get the updates setup so they prompt you to install and make time to get it done. Don't keep putting it off.

Task two make sure that you have a good antivirus solution. I would recommend you pay for one like Trend Micro, Symantec, Malwarebytes you choose but make sure you consider a paid solution they are just better than the free versions as you could understand. Why would a company put all of the functionality in free platforms, they don't make any money from them, that's why they have a paid version. You get the best version of their platform. The free versions are only supposed to be an example of their platforms, not a permanent solution.

If however, you can't afford to use a paid antivirus solution at least use the free AV and not run without any protection at all, that is just asking for trouble. Would you just go out and wave a red flag at thousands of vicious, scary-looking bulls? No, you wouldn't, so why would you have no AV on your computer, It is just as dangerous. Okay, maybe not quite as dangerous as thousands of angry bulls but you get what I am saying right. Don't wave your red flag enticing the malicious actors to come and knock your virtual door down. Secure your systems up with at least basic protection.

Task three ensure that your home Wi-Fi connection is secure so that no one can connect to your network without your knowledge or at least reduce the likely hood of it occurring. If you can do it yourself I would ensure that it is WPA2 at a minimum. You need to also ensure that WPS is turned off on your network so that you don't make it too easy for someone to break in.

If you are a little technically capable or are getting someone who is to help you do this I would also consider locking down the Wi-Fi to only allow specific devices. This can be done by entering the MAC address (physical hardware identifier that all devices that have any network interactions). This will mean that if the MAC has not been approved it will not be allowed access to your network. This is perfect if you only have particular devices that are used, however, if you allow your friends or family to connect up and use your internet you may not want to add this because of the issues it could cause. It would be good security protection if you don't need to allow random devices though, so at least consider it.

Why you or whoever is assisting you to configure the Wi-Fi is working on the device, you should get them to verify that your router/Wi-Fi access points are all up to date with the latest firmware (basically the devices internal software). If it is not on the latest version, do it immediately it will make the device run better as it fixes issues and will improve security as vulnerabilities will have been fixed.

Task four you should now start to look at how you create and store your passwords. Many of you will use the same passwords on all or most of your accounts. This is bad, please stop doing that. I know you hate trying to remember all of these passwords and they all want you to have random numbers and letters and symbols. I get it, it's hard. How about when you are at home you use what is called a password manager. You can create a nice and long passphrase to ensure that the password manager is secure then you can have that password manager generate nice and long individual passwords for all of the different sites or accounts you have. The passwords are all encrypted and secure but if one online account is breached you won't give them access to all of your accounts.

You need to ensure that the password manager does not stay logged in when you close your browser, this will ensure that each time you want to go on the internet it will need to ask you for your main passphrase but it will ensure you keep the accounts secure from anyone who gains access to your computer. You can never be too safe.

The above four items are in no way all you can do to help stay safe online but if you can get the minimum covered and we work from there, we will all be much safer for it moving forward. Do yourself a favour and take my advice, you won't regret it.

Now if you are feeling adventurous you could also consider implementing a good base model firewall and IDS platform. This is probably way above what most people would need or want but if you do want to go that extra mile it could be worth giving your home network security a shot of steroids in the arm. Yes, it will cost a little but probably not as much as you think. Talk to your work, if you do a lot of work from home they may be willing to subsidise your security improvements as it will help keep their systems safe as well and in comparison to the corporate network adding an extra licence would be a trivial cost increase.

Yes, they could say no to you that it's not their responsibility but you don't know unless you ask. They might even allow your internal IT support team to help you get it all configured nice and securely that would be a win, win don't you think?

41

THE ARROGANCE ISSUE IN CYBERSECURITY

This is a topic that I have considered writing about for quite a while but I wasn't sure how I should approach it. So for this chapter, I am not going to think about it too much and just write what I feel (it could be risky, maybe even disastrous but here goes). I have been interacting and deepening my involvement with the security industry since 2013 (7-8 Years) which admittedly still makes me feel like a relative newcomer, especially since it has only been 3 years since I was employed in a cybersecurity only position (starting with Davichi 2017), yes I have been doing security as a part of my job for 8 years but even that seems like a short time when you look at some others in the industry with 20-25 year's experience.

I have been in IT since the early 2000s which is 20 years which I feel gives me a little more credibility but some in the industry don't see it that way. To them, if you haven't been in the industry for 10-15 years you don't know what you are talking about and I want to say that I think that is a load of crap. I think that I could learn something from the internet who has just started on their journey as much as I could from someone who has been in the industry since the dinosaurs roamed (yes I know there was no computers when dinosaurs were around but you get my point).

Especially in an industry such as ours, things change so fast that if you blink for too long you can miss something truly significant. I am not being dramatic here, I am being serious. If you don't have your finger on the pulse you will get left behind which is why I think the old school way of thinking just doesn't cut it anymore.

If you are one of those people that are arrogant and act as though they are better than everyone else. Who will not share their knowledge with anyone trying to learn and gain access into this awesome security industry of ours, then I am sorry but your time is numbered. You will become obsolete and fall behind the upcoming entrants, even those who have been here for a few years. No one will look up to you anymore and no one will put in the time to help you stay current because you wouldn't invest in them, why in their right minds would they waste any time at all on you.

I have dedicated myself to learning this mysterious craft we call cybersecurity and have seen so many others do the same. We share our knowledge any chance we get and we don't act like entitled rock stars. We are just professionals in our crafts who want to make it a little easier for the next generation, teach them about mistakes or stumbles we had so they can try to avoid them so that in the end they become better than us. Just like parents do with their children we want to pass down our knowledge and help that next generation reach new heights.

If you are joining the industry because you want to be a rock star turn around and go back to whatever place thinks that is cool now because security isn't for you anymore. We want collaborators, artists and phycologists. We need people who are good with people, we don't want to hide in the shadows and practice our dark magic looking all mysterious. That is not who we are as an industry anymore or at least that is not who we need to be.

I know that this chapter is starting to seem like a bit of a rant which I guess it is to a point but I need to get this point across clearly, no fluff involved. The old school arrogant security persona needs to be extinct, we need everyone to stand together and help organisations and individuals do better with security. Up until now, we have been on the back foot and we are losing this virtual battle.

The bad guys are winning it's that simple.

So, how about we start a new chapter tomorrow in the industry, how about everyone gets up as you would normally do and go to work (if you aren't isolated to that COVID-19 virus that is taking over the world as I write this). Today however look for ways you could help someone entering the industry, answer an email or point them to a resource that helped you. Have a coffee with someone that has reached out to you asking for advice.

There are so many ways that we all can do our part and for the love of god stop the arrogance, we are better than everyone else attitude. You may be more educated, you may have been around for 20 years but that doesn't mean you are better than everyone else. Let's step off our pedestals before we fall, we are all just humans after all trying to make a difference in our own worlds.

42

DO YOU KNOW HOW VALUABLE YOUR MOBILE DEVICE IS?

As I start to write this chapter I am looking down at my smartphone and it concerns me exactly how much of my information is on this. My whole life is on it, emails, contacts, photos, Multifactor apps, even my travel history (yes all our GPS data is being tracked by our devices, they know everywhere we have been). I am not alone in this though, think about what you have on your phone?

Cached logins, cloud data storage, work/personal emails, pictures of our children (some people may even have a bit more risqué stuff – Trust me it's pretty common if this is you delete the images or videos before malicious actors steal them all and share them with your grandparents). Honestly, I hear about instances of this all the time, just keep in mind if it is available on your devices or online in any form you are playing with fire. It is only a matter of time before some nice malicious actor or just some hacker that you rubbed the wrong way to get on and make everything you would like to keep privately available for everyone to see. I have told you so I guess you can do with that what you will.

So with that being said who backs up their phones regularly, honestly I wish I could say I did but I don't think I have done it for months which is probably an issue I should quickly resolve but I am almost certain a large percentage of people who read this book will not backup t their phones either. Most don't have any antivirus solution or data protection solution on the device (Now that I can say I do have) but with so much data on them and with how bad a malicious actor could make our lives if they had access to it all, shouldn't we do more to protect them or at least think about the threat a bit more.

Someone at your home or office could be walking around near you right now with an infected phone. With the Emotet virus now capable of spreading itself through open Wi-Fi connections you need to make sure that you are being more secure.

Let's go down one of my scenario rabbit holes for a moment. You are an hr team member with a large multinational organisation with more than 10,000 employees. You have access to financial and personal information to all of these staff members and higher up managers. You regularly receive email communications from staff via email and through the HR portal which you have an app on your phone for access as you need it. You have a shared cloud storage location with Box/OneDrive/dropbox (choose one) with access on your mobile devices for ease of access and the ability to email or upload files to the HR portal or review other team members completed documentation when you need it.

You do all your banking via your phone as it is the easiest way to do things and you generally don't have time at work or once you get home so mobile banking is the best way for you to manage your accounts. You have google authenticator, Microsoft authenticator and some vendor-specific MFA apps (this list is growing) as well as some of your more personal apps like Facebook, Instagram, LinkedIn and Tik Tok.

You maybe even have a few rideshare apps like uber and didi as well as domino's pizza. All normal stuff. These are things you would find on most people's smartphones nowadays. You go out for lunch with some friends (if you still can, I am writing this during the whole COVID-19 pandemic that has hit most of the world and is starting to cause full lockdowns here in Australia). You are having a great time, laughing, eating and having a couple of drinks. After lunch, you head back to the office and get back to work. After a few hours, you reach into your bag to find your mobile (time for your afternoon socials check) but you can't find it. You conclude that you must have left it at the restaurant but when you call they say that no phone was found or handed in. you are hoping that it turns up but for the moment you need to get back to work.

That afternoon you get home and you decide to jump on to pay some bills on your home computer now that your phone is missing. Your bank account has been drained of funds. You start to freak out a little, someone must have my phone. Maybe you should have used a better pin than the default 0000 shouldn't you? You check your credit cards and there have been multiple purchases and your cards are maxed out. Oh, this is going to be a bad week.

Suddenly your home phone is calling, you see that it is the office. On the phone is one of the support team who has had one of the other team reports that the cloud storage has been deleted with nothing left but not before it was all copied off by an unknown party. They have traced it back to your account that was used to do so. You inform them your phone is missing and that you had also lost funds in your bank accounts. They indicate that you should reset all passwords and call your bank.

This is a bad scenario but very possible. You need to ensure that you know your risks with your phones and take steps to ensure that they are safe and secure. Put antivirus on your devices, don't cache the logins -make sure you need to enter every time. Put a pin that is not 0000, don't use anyone's birthday or something just as easy for a malicious actor to guess or even just an opportunistic person who came across your lost device. Don't make it too easy for them.

Ensure that you or your company can remote wipe your phone the moment it comes online, logged in or not. That is essential to ensure data is protected and allows a little bit of peace of mind if you do misplace a phone. Yes, it may be annoying to have to reconfigure a device but if you have a backup (which I know we all do right? Mmmm of course we do) then it will be a quick restore process if it is found. Trust me it's for the best, get your backups in order.

I know we don't always consider our mobile phones when ensuring we are prepared for any incidents (myself included) but we all need to just do the groundwork, pay for a secure cloud backup and ensure you have the correct security measures in place.

Oh and one more very important thing, stop jailbreaking your devices so you can play some bloody game. STOP installing apps that are not from a genuine app store. This is something that can help to keep your devices clean and safe. Not from being lost or stolen but from being infected by some malicious software and used as a zombie in some botnet or just all your money and info stolen.

It's simple, be smart and don't do something that will put you and your devices at risk.

43

SOFT SKILLS THAT ARE NEEDED IN CYBERSECURITY

This is a topic I seem to have conversations about with almost every person wanting to get into cybersecurity and almost every recruiter in the space. However, even though it is a conversation I have a lot, I don't think that most people see the benefit of what I would say is really useful soft skills. In some cases, these skills would be more valuable than technical knowledge or abilities. I have always said that technical skills can be taught but soft skills are a much harder skill set to pass on.

Why don't we explore this a bit further and maybe you will agree with me, maybe you won't but I think it is a really important topic that I want to discuss, no better place than as a chapter topic for my second book right? That's what I thought too.

Soft skills come in many forms, some that personally feel would be beneficial for someone who is looking to enter the cybersecurity industry would be:

- Curiosity
- Perseverance or a dog at a bone mentality
- Ability to connect to other people
- Ability to put themselves in another's shoes/mindset
- Drive to continue to develop themselves

The attributes I have listed above are just some of the many useful skills that I would look for in a potential entry-level employee. Let me go through my reasoning behind some of these and afterwards I feel you will understand where I am coming from. Maybe you may come around to my way of thinking and you could help someone join our industry.

Curiosity is a big one for me, I believe that in security you need someone who naturally wants to know more about something, wants to get in under the hood of a virus or incident and want to get a better understanding of what has occurred. This can be something that will entice the person to follow the breadcrumbs of an incident and help them piece the puzzle together. Learning and developing that natural curiosity will make that person no matter their current technical abilities a valuable member of an incident response team or a security researcher.

Perseverance or dog at a bone mentality. This is a key personality or soft skill that can make you a good or bad security professional. Why? I think it is very simple. Let's say you have just been breached and you are looking for evidence of the incident with nothing at all to be found. You need to have a security professional that will sit and go through a mountain of logs, review every detail and I mean every detail. This could take weeks or months to go through everything, put all the pieces together. This is what we need more of, not people out to get quick fixes. We need these people who will just get a slight sniff of something and just keep scratching at the wall until they break through and put all the pieces in a line. They won't give up after a few setbacks, it just isn't in their natures, it would just gnaw at them until they go back and keep trying.

That kind of perseverance is essential and something that I have never been asked about personally but feels that hiring managers should take more value in this skill.

Ability to connect with other people. Being able to truly talk to other humans and make real connections with them is a skill set that many technical people are completely missing, many skilled technical people hate human interaction, they don't need or want to have that kind of interaction. An introvert is normal. I am a bit of an introvert but over the years have taught myself to break out of that mould. To break from my mental restraints for the better. This has given me a unique mindset and ability to understand most technical things and also be able to translate, relay this information to people around me that are not a technical person. It is a skill that I have continuously honed and will continue to do so.

The ability to talk to people about complicated things and help them understand is something that any entrant would be very successful if they could master. This is hard to teach but for someone who is naturally wired that way which I am (with some encouragement of my introverted self). They can help build awareness in organisations, build buy-in with staff through awareness programs and help turn the tide of incidents around. Humans are the key to us winning this cyberwar we are all waging, not all of us know it yet but it is true.

Some still think that blinky lights are the key but that is not the case in my opinion. Most breaches are not some fancy zero-day attacks, they are unpatched systems or humans are exploited through phishing, phone or text scams, in-person manipulation and that list goes on. Humans are a massive part of the malicious actors attack vectors. If there were no humans, they would have to rethink their whole process, they would have nothing and they would all go broke with revenue drying up faster than a glass of milk disappears with a ghost chilli being consumed.

That chilli scares me, crazy hot and I am never going to be eating one of them. But it demonstrates my point. My point is valid though if we can bring in more people who can connect and relate to people we will have a much easier job as defenders. People might start to like the weird dark magic conjurers that have been hiding in the office basement for years sacrificing goats to the cyber gods. Seriously we don't do that as hackers or security professionals, it's not a thing. If someone somewhere is, they are alone in that weird stuff, that is not an industry thing.

The ability to put themselves in another's shoes/ mindset is part of the human interaction if you as a security person can consider a scenario in the way of your attacker or look at scenarios of a potential victim you can adapt your processes or configurations to adapt for these possibilities. Know your enemy and know how they work, how they attack. If you can get into the mindset and understand why someone has done something one way them you will be able to predict the next moves, react before they make it.

Breaches will be better handled and you will be a very valuable asset to any security team. I have always said we should bring people in from different industries, psychology is one of the very useful skill sets that will become more and more useful as we continue to evolve as an industry. They will be able to truly understand the opponent. That will be an enviable skill that I can tell you.

Lastly, the drive to continue to develop themselves, could be one of the most important. People in our industry need to continue to always evolve, always strive to continue to stay at the forefront of techniques and attacks otherwise they will become obsolete like the dinosaurs of old. We need to continue to learn and study and read. Never stop, have a thirst for continual learning and self-development. I know myself, it's like I truly never quench that thirst for knowledge. I have completed two masters' degrees and still think I might do more.

It doesn't always mean formal study either, it could just mean keeping your finger on the pulse so to speak. Hunting for knowledge in the deep web, learning what the enemy is doing, what is valuable for them now and what has lost its flavour. These things can help direct your efforts at the right issues, not ones that don't matter anymore. When the malicious actors change tactic we need to know as quickly as we can to reduce the effect or costs.

Thirst for knowledge is key to this. If you want to always keep pushing it will allow you to succeed and could help you provide that missing piece we as an industry have been looking for.

So, can you now see why some of these soft skills are so important? There is more but I think you get what I am trying to demonstrate, let's look outside the box, let's stop trying to put square pegs in round holes. Let's forget the cookie-cutter professionals we are all trying to find (the unicorns) and let us bring in the people that in years to come will surprise us all by how much they can truly give to the industry as a whole. I know we will think back on this and go say why did we push back on this for so long.

Go, find your unique individuals who have the soft skills that you need and teach them everything else. They will be loyal and in the end, they may be the best person on your team if you give them that chance to prove to everyone they belong.

44

NON-TRADITIONAL PATHWAYS INTO CYBERSECURITY

Cybersecurity is a tough industry to break into, I know from experience and I have had many conversations around diversity, soft skills and non-traditional pathways in my articles/podcasts/webinars you name it I have tried to discuss it on so many occasions. Why? I think that to truly succeed in this great industry of ours, we need to stop looking for technical people (okay not completely that would be insane) and start to look at different fields for new participants. Let's stretch the idea of who can be a useful addition to cybersecurity.

The industry to my knowledge is filled with mostly technical backgrounds like mine. I have been in IT since I finished school way back in 1999. Wow, that feels like so long ago now. I spent about ten years doing many different roles in IT before I started to move over to cybersecurity and during my discussions with other members of our industry that is a similar path that many took. IT transitioning into cybersecurity. That transition makes sense but what about a lawyer, a psychologist, a doctor, a public speaker or even a biologist?

I think we could use some more people from all of these backgrounds and more. I have seen people make a move over to the space with great success. Nicole Stephensen – Ground Up Consulting is one of these people. Yes, she is still doing legal work but has focused on the privacy area which has enabled her to join the cybersecurity industry with the backing of her traditional background. There are many more stories like this and I would like to see them more.

If you are a psychologist, for instance, I believe that you would be able to bring to the industry a great ability to understand how someone thinks, get a real insight into why someone has done something or reacted in a certain way. I can only imagine what it would be like to truly dive deep into the mind of a malicious actor, to get in deep and know how they tick, what makes them who they are. That kind of insight and knowledge could make a massive difference in defending an organisation or even help to generate threat intelligence for the industry.

Public speakers and even politicians (I know it's a stretch but some are very smart people) could be great at helping the industry connect with everyone who is not part of our world. Think about it, security awareness is not working as we need it to yet, too many users/staff are not learning and improving their skills. They are still vulnerable to threats. Yes, nothing we do will change that but we need to do better, find new ways of challenging this issue. Find out how to connect and teach people how to protect themselves better.

How about police officers, they would already have an understanding of how criminals behave, crime is just crime after all no matter the platform or environment. Yes in cyberspace there would seem to be less risk of harm for criminals but the insight from a policing background could be very helpful in intelligence and cybersecurity as a whole.

I am almost certain that if I tried I could probably find a reason why almost any profession could meld with cybersecurity and help make that person a useful member of our industry. I think that you can see where I am going with this, you don't need a technical background to come to join this awesome industry of ours, yes it is sometimes very challenging but if that is something you will enjoy we could certainly use the variety of skills to help us defend our systems.

So let's do what we can to encourage true diversity in our industry and let's make this industry of ours even better.

45

NOT A HOLLYWOOD RED TEAM STORY – HOW IT IS REALLY LIKE

I have been thinking about this chapter for a while, I have been asked on a few occasions what it is like to be a Pentester or ethical hacker. I know everyone wants to hear about some sort of James bond style hacking incident where I was running some spectacular attack against a challenging opponent that just wouldn't submit to my command. I know you all want to hear about some awesome breach that I was involved in during a red team job but honestly I have been thinking about it hard and I cannot think of one gig in which I could tell that would be worthy of a Hollywood story.

I could make one up but I don't want to do that, I want to tell it how it is. So with that in mind, I am going to tell you all how it normally is on a real-world pentest job. Nothing magical or fantastical just how it really would be if you decide to become a Pentester or red team member in the cybersecurity field.

I will do my best to not make it too boring but run you through some of the stages or steps you would likely take from start to finish and try to pass on how I would approach something during the process. I apologise in advance if you were hoping for a James bond style story but for you check out the end chapters of the book. I will be providing you all the first 10 chapters of the new hacker fantasy I am writing "Foresight". I look forward to hearing what people think of the fantasy section and will work on getting the completed book in all your hands by the end of 2020 at the latest (Fingers crossed).

Back to our topic at hand. Let's create a hypothetical client to do a pentest engagement with, let's call them Moneybag industries, just for entertainment sake and let's say they are a large corporate accounting firm that has clients and offices all over the world. They have more than 1000 staff geographically dispersed with multiple help desk teams that are also spread around to allow for 24/7 support following the sun with different locations starting as others finish. It's a common scenario. They have asked us to conduct a full security pentest and social engineering attack against their systems/users. They want to know what the weaknesses are. We will be starting an assignment from what I would call a cold start. That means that we have no systems information and will be required to do exactly as a malicious actor would and gather all information about the company and who works there. What systems they use and anything else that will help us gain access to the systems.

Step 1 – information gathering – non-touch. In this we will research the target company without doing anything to raise any red flags or notify anyone that someone is looking into them, gathering intelligence. I would normally use social network platforms to find staff, collect information on any that could be a useful target (a user who may have elevated permissions), look over job add sites to see if the company had recently advertised for any IT support positions. This would give us two things, insight into the platforms being used by the company because these will be likely listed as skills requirements or certification likes. It would also allow us to know if we could use the new IT support person angle for social engineering against the new hirer or unsuspecting users who don't know all of the support team because they change too often.

Step 2 – Analyse information. At this point, you would need to review all of the information you have collected and determine what is relevant and what isn't. You would then determine from what you have what further information gathering you would need to complete to fill any gaps in the information you have on your target. At this point, you may use some soft-touch methods by looking over the sites the company has, do some stealth scans on systems so that your presence is not detected. You need to be careful in this stage to ensure that you don't do anything aggressive or obvious to raise any flags. If you are detected now it will make your job very difficult for the rest of this engagement. So be careful.

Step 3 – Potential exploits. Now that you know more about what platforms the target uses you can gather an arsenal of potential exploits. You may also be able to consider non-technical attacks such as social engineering or even scope out physical site locations. If it is feasible you (and part of your job scope – you don't want to go to jail for something that is not part of scope no matter how fun it sounds) could consider some methods of attack that sees you attend an office and just make your way in and find a way to gain the access you need to systems.

You need to come up with a plan of attack, remember in pentesting or red team engagements you will need to stay flexible and go with the flow of attack, adapt your plan for what is found and always stick within the allowed scope (I know it can be painful and not much fun but it is better than a 3x3 cell with bars and no creature comforts).

Make sure when you are going to attack, you are ready for anything that may come your way. This will help you to act quickly and reduce the chances of being caught.

Step 4 – Launch of attack virtually and physically. This is the fun part of most engagements, in my opinion, you get to try out exploits and dig in and get your hands dirty. You will throw your hammer at the defences and see what you can do, try exploits that match systems and carefully manipulate your way in (more careful than a malicious actor, they don't need to explain what the hell happened if they crash a customer's systems, you do).

This stage could go on for days and days before you have exhausted all opportunities, maybe even longer but you need to not get frustrated with the process, keep calm and don't do anything stupid to give yourself away. Once you have exhausted all opportunities whether you have gained access or not at this point it is time to switch approach.

Step 5 – onsite and social engineering attacks. I regularly utilise these methods in engagements but they always make me slightly nervous. No one wants to go to jail so I always ensure that I stay within my boundaries of job scope. Once you know what is and isn't allowed be creative, look outside the box to figure out how you could gain access, people are normally quite willing to help you just need to ensure whatever you decide to go with it doesn't throw up red flags, you need people to believe that you are who you say you are and by whatever it is that you are selling them.

Some people are better at this than others, I am honestly not the best at this but I am getting better over time. I am best at finding ways into building or offices and making it look like I belong there. I regularly walk in and set myself up in a meeting room or board room to work. I even ask directions sometimes, while dropping someone's name that makes sense. It works sometimes and sometimes it doesn't but you need to remember that if caught out the gig is essentially over. The rest of your work doesn't require much stealth just good old fashioned time to analyse and review.

Step 6 – finalising and verifying any access or vulnerabilities found. This is the part of the testing most pentesters or red teamers start to turn off a bit but it is important to ensure that your work and results are valid. You don't want to look like a fool if what you say is a vulnerability is actually a false positive or maybe even you being tripped up by the honey pot or something. That would be awkward.

Step 7 – Reporting and client discussions/debrief. This is one of the most important parts, you need to be able to help clients understand what you did and what you found. You need to be able to give them clear direction on how to resolve the issues found, no you don't need to fix it but you should guide them on what will be needed to ensure the security holes are closed up and give them any further recommendations that could be useful moving forward.

That is now basically the main steps that a pentest or red team engagement would follow this is just an overview and as a Pentester you would need to ensure that you understand each step and what is required. As I indicated at the beginning of the chapter this is far from a Hollywood drama about red teaming but a basic insight on what it would be like if you joined the red teamers on a job.

I get told by a lot of people that this is the area of expertise they want to get into but there are limited positions in this area, so consider not narrowing your sights and look at awareness training or incident response. I know they are not as sexy roles but they will still be helpful and quite interesting careers. Just want you all to consider all your options and not get fixated on one area to have that blocked for you because of the lack of skills, ability or just no opportunities.

46

FINAL THOUGHTS TO LEAVE YOU WITH FOR A HACKER I AM – VOL 2

The cybersecurity industry is a great community and I am glad I can say that I have truly become a participating member, not only learning from others but also being an active contributor. I am grateful to everyone who has helped me gain my understanding of the community and has helped make me a true member, not just a bystander.

I am writing this chapter on April 1st 2020 and I am reflecting on what has been a tuff month for everyone. At the beginning of March, I made a post on my socials that I was looking forward to March, I was going to be doing a book signing at the Brissec20 security conference and was going to be doing a webinar session with Charles Sturt University and IT Masters about my experiences with doing two masters degrees and the personal challenges I had getting into cybersecurity.

I did the webinar with CSU and IT Masters on the 16th March and I thought it was great, the feedback was good from attendees which is all we can ask for. I also appeared on "The Get Cyber Resilient Show" podcast with host Garrett O'Hara from Mimecast, who I mentioned in one of my other chapters, this was a first for me for both situations, yes I had done in person talks at conferences and I regularly conduct cybersecurity awareness training for our clients but this was great fun and I honestly hope I get to do it again with both sometime soon.

That was the only good part of the month sadly though, I chose to step down from my CSO journalist position at the beginning of the month (for reasons I won't get into) and then the COVID-19 virus pandemic arrived, the whole world went into an almost lockdown (still is while I am writing this chapter), all local clubs and pubs were shut down and most other non-essential businesses that couldn't work remotely. All security conferences were cancelled/postponed till further notice which means I now have nearly 20 copies of my "A Hacker, I Am" book for no reason, I guess whenever Brissec20 is back on I will be prepared for the book signing, will give me time to get some copies of this book (A Hacker, I Am – Vol 2) for the event as well.

I was also sadly made redundant from my day job due to an unforeseeable crunch on the business due to COVID-19, I have a couple of weeks left before I finish up and I am truly saddened by the whole situation. So many people everywhere have lost their jobs and it is only going to get worse before it gets any better. I hope this book sells well, then I may not need a regular job anymore. I very much doubt that will happen anytime soon but I can always dream, I just don't see any flashy new red Ferraris in my future.

So, as many others are at this time I will join the job hunt queue, it's certainly not something I have needed to worry about for a long time but I will embrace it, that's all we can do.

The cybersecurity industry is in high demand right now though, so we all need to stay vigilant. If we are working or not we have the right skills to help others keep the suddenly all remote workforce secure and protected at home. Help them all make better decisions and help make them work well (keeping the ones still working in jobs).

Malicious actors are trying to take advantage of this situation, one in which no planning or at least very minimal planning has converted companies with no remote access, now wants everyone to do it and have it running in a couple of days. Smells like trouble, don't you think? Most companies are not ready for this magnitude of change, they don't have the security in place to ensure that staff and business systems are safe.

We as the industry need to stand together and help them bring their systems up to scratch, help them get at least the minimum right. Let's do what we can pro bono, help charities, help businesses in pain that just need that helping hand to keep afloat, that way when we get on the other side of this mess we will still have jobs and businesses to protect.

So for my final thought for my second book "A Hacker, I Am – Vol 2" let us stand as one and let's truly make a difference together, let's forget we are competitors just this once and let's just do what's right. We will all be stronger because of it.

See you all on the other side in the new world, whatever that is going to look like.

Chapter 47 through to 56 are different. For these next ten chapters, you will get a sneak peek into my upcoming hacker fantasy novel "Foresight" I hope you enjoy the first look at the book.

Foresight

47

FORESIGHT: IN THE BEGINNING

Life is funny isn't it, you think you know what you have instore but then something happens and it changes everything. Sometimes for good, sometimes for the worse but it's those moments that seem to just happen in front of our eyes without us even noticing or acknowledging the occurrence but the impact of these events are far reached and can change the entire direction we were working towards. I have had several of these moments in my life but none more prevalent in my mind than losing my mother.

Now there was no gruesome murder or horrible car crash or anything along those lines. My mother just woke up one day and decided that she didn't want to live the suburban lifestyle anymore. She didn't want to give up her youth being a mother, she was only 18 years old when she had me. My parents were quite young when I decided to grace the world with my beloved charm and throw their plans to the wayside. Yes, I am almost one hundred per cent certain that I "Samantha Erkhart" was a mistake, surprise, nightmare and so totally unplanned that I do not doubt. My parents were almost finished high school when Jenny (my mother) found out that she was expecting a child. She cried for days when she found out if you are to believe what my grandparents told me when I asked them once when I was about 11 years old.

Jenny had plans, finish high school, go to university and then travel Europe for six or more months before coming back to make it as a big-time lawyer or accountant or whatever it was that she thought was more important than being a mother to me. I don't really remember her much I was only two when she left, I just know she was very pretty from the pictures I have seen of her, my grandparents occasionally pull some out for me to look at but I assume she has lost some of that beauty by now being that it was 15 years passed, since she walked out on us. I hope the years haven't been kind to her and they have taken their toll. I hope that she failed at achieving the glorious life she walked out on us to create, I hope it haunts her every day what she did to me but I bet she doesn't lose one ounce of sleep and never has. Jenny is a self-centred bitch and I doubt she even cares about how I turned out all these years later.

I think it's pretty obvious that I am not a big fan of my mother, I am a little bitter and resentful. I think it is part of my charm. The whole moody teenager fits perfect for my stereotype upbringing don't you think. I have had a pretty good life with my dad (John) though, I honestly have and I think we have been better off without Jenny. I know my dad misses her sometimes he surprisingly still loves her and never says a bad thing about her at all. If I were him I would trash talk her till the cows come home, she left us with not a second thought so why should we give her the satisfaction of thinking about her.

He has never gone on any dates or even really looked at another woman since she left which is a little sad for my dad but when I have talked about it with him over the last few years about him putting himself out there, he always just says that he has everything he needs with the two of us. I love it when he says that but I would never admit that to him, maybe I should though.

John (Dad) works in construction and has done for as long as I remember. These days he manages builds of big skyscrapers and they keep him very busy. So busy that we don't see much of each other during the week but he always comes and says hello when he gets home at night. I love my dad, he has always made sure I was looked after and I know on a few occasions he went without a lot of things to ensure I had a roof over my head, clothed and was fed well. A great dad. I hit the jackpot on the dad front that's something I know for sure.

Dad not being around though has allowed me to get entangled in a world he wouldn't even understand or probably approve of. A world of deceit, manipulation and a lot of filth if you knew where to look for it. I am a hacker, yes a hacker. I laugh when I see the depiction of a hacker in news or media, they are mostly men in dark colour hoodies with this weird background of "matrix" code streaming down the screens. Honestly, a large number of the best hackers in the world are girls just like me.

I look more like the girl next door than what everyone thinks hackers look like. That misconception though can have its advantages for me, who would expect the girl next door? No one right? They would believe that I would be more likely to be out getting my nails done than be able to hack into your car's autopilot systems. I saw a couple of guys win a new tesla for just pointing out basic issues but not sure if I should tell them I could take full control of the car without interaction from the occupants, I did that to our neighbour's car just for fun. He's a perv, I see him looking at me all the time and I just wanted to have some fun with him maybe teach him a bit of a lesson. He doesn't know about that though and I should probably try to keep it that way. If only he knew. Makes me smile just the thought of how safe people think their electronic devices are, I think many would lose a lot of sleep if they knew what some of us can do.

People have no idea that hackers are just normal people, you probably know some of them but you just don't know it, they are probably right in front of you. I have been one of them for three or four years now maybe longer. If I really think about it, I have probably been one all my life. I have always just had this natural ability, almost like machines talk to me. When I have a target, I kind of just focus and I can see what I need to do. It sort of visualises for me in my mind's eye or something. I know right, it sounds crazy just thinking about it but it's a gift that has made me the perfect hacker.

I have done some things to become part of the community in the underground scene where I am known as **Foresight**. I know it's a bit corny but I visualise my attacks and I came up with the handle when I was like thirteen or fourteen years old. It was about then when I started to poke around the deep web. I always heard of this dark festering cesspool of a place called the dark web and I wanted to know more but mostly it's pretty tame. It's just a bunch of old conspiracy dudes not wanting their governments to see what they do online, oh and then there is the usual terrorists or criminals selling everything from your baggie of marijuana to warheads. It is a place you can buy whatever you want, no questions asked as long as you have enough bitcoin to pay for it.

My normal world is pretty average, although I don't know if the world everyone sees is my real-world or if the life I hide from everyone in cyberspace is my truest self. However, let's just say that the 17-year-old high school student Sam is my normal world. I am a normal 17-year-old girl, I want what anyone wants, to get through high school reasonably unshaved. I am not one of those popular girls or one that is very popular with the boys for all the wrong reasons. I just go to school, do my work and just let it pass me by just kind of watching it all from a distance.

I have my friends and school isn't that bad but there is nothing extraordinary about this existence. I am just an average girl next door, Yeah I am reasonably pretty and I get noticed by the boys on occasion but that doesn't fascinate me. I shut down any sort of advancement quickly, I just want to finish my time here and move on to college. I want to break the girl next door image and become something else. I am not sure exactly what that something else is just yet but there has to be more than just this for me in my life.

I need to get out of my head and finish my homework before John gets home, it is Friday and we always have dinner together in front of the TV or what I like to call the idiot box. We always watch some sort of new movie together, we don't have a particular genre or anything. John just chooses one at the DVD store on the way home from work and gets us a burger and fries from the takeaway store next door. I could probably get all of the movies for free but I don't have the heart to tell my dad that. It's our thing and I like it.

It's strange how little things like that matter, isn't it. Just that time we spend together each week means more to me than anything else in my life. I think it does to my dad as well but I don't think it would be something he would admit though. It would be a little too mushy for his liking. I hear his truck pull up in the garage and it puts a smile on my face, one of those real smiles that comes from the inside not one of those fake ones many of us put on when we greet someone on the street or something like that. I wonder what corny movie he has chosen for us today, I hope he didn't forget about the extra salt on my chips. I am a bit of a fan of salt.

I reach over and close the lid of my laptop and make my way down the hallway, my dad is unpacking the takeaway and setting it out on plates for us to go sit in the lounge room together. As I pick up my plate I see he has opted for a favourite of mine for video choice tonight – The matrix. It's a little funny really when you think about it that neo would be an idol of mine, it's a wonder I didn't make my hacker handle neo. I feel that would probably have been too much though. Tonight should be a good night, I just have a feeling about it.

48

FORESIGHT: AFTER SCHOOL ACTIVITIES

Like many teenagers school is not my thing, don't get me wrong I don't hate school or anything but it just goes by as a sort of daydream that I am taking part in but don't engage with. It just happens and I go with it. Like the flow of a gentle river, it just turns and dips just going in the direction it is encouraged to go but it doesn't interact or respond it just flows. That's how my days are in school I do as required, interact as little as possible and just wait. Wait till the day will end and I can re-enter what feels more like my true reality. Cyberspace, the internet on all levels, not just the ones most see but the deep dark corners where the boogie man would hide all his secrets.

I feel really at home as Foresight, my alter ego of sorts. She isn't like me though, she is ruthless, unforgiving and is a respected member of the hacking world. Maybe even feared a little if I am being honest. I have been working on a project of sorts the last few weeks that is bigger than I have ever done before, it will concrete my place in the cyber world if anyone ever found out it was me but it is risky and I need to make sure that I am taking the proper precautions. It's not going to be easy.

I have already gained access to all of the primary sites the marketplace is hosted from and have escalated my permissions. My payload has trickled down through into all corners of their network. Every machine in every location has my agent on it. It was a bit of a labour of love, the agent, it has been created to shut out all other interactions or commands and only accept mine. It has a self-destruct component that if I do not send a pulse within 24 hours after the assault starts, it will start an erasure protocol and hide my tracks for me. Although I have been careful and cleaned my tracks as I moved through the systems. Ensuring that I only clear my logs and my activities. All others remain intact.

I know what you are thinking, if I created an erasure protocol in my agent then why clean my tracks as I went. Simple, if my agent is found I don't want it to come back to me in any way, not just for my safety but for Johns as well. This will not just come back on me but Dad will be in the firing line. These people are truly the worst kind, the real scum of the earth, if I am careless and leave some sort of way to connect it back to me the consequences will be life or death.

A mistake with this will result in me turning up in pieces somewhere or taken and added into their prostitution ring forced to do ungodly things that make a shiver run down my spine, just thinking about it. If I don't do something though, I am just squandering my gifts, wasting my abilities. Today is the day I use them for something greater than myself, I know I have taken precautions and it won't come back on me but I understand my fear. I need to use that fear to fuel me, not to hinder me.

Burrrrt, the school bell sounds. It's time to start my walk home, it's about a 30-minute casual walk home but I have a feeling that today won't take that long. I pick up my books and shove them into my backpack and head directly for the door. I start to map out my path through the hallway and down the stairs before heading out the front foyer area, down the front path leading out the front gate of the school. I nearly made it all the way too, when a senior boy put his arm out slightly in front of me just inside the gate.

"You're Sam aren't you?" I nod slightly. "I'm told you can get things for people?" I look around to see who is around and don't see anything out of the norm. I look back at him and nod slowly. Watching his every move, trying to see a sign if something is off. He reaches into his pocket and pulls out a handwritten note. Fortnight super health and weapon upgrade will give you $30. I consider for a moment, I am not a gamer at all but I know this is the current "It" game that everyone wants, I don't see the point, 100 people dropped on an island to kill each other off until the last one standing wins. It's pretty popular though that's all most of the boys talk about and even some of the girls. I think the reason I am not interested in it is that it wouldn't be much of a challenge. I look up at him and just say "I will need your log in name and the fee is $50 in advance". He looks at me as though he was considering arguing but something in the look on my face must have deterred him and he reaches into his wallet and takes out a $50 note. He scribbled down his gamer login name on the bottom of the piece of paper he had previously handed me and then folded the money and paper together before passing it back to me. I step slightly to the side and walk out the gate.

As I walk down the front the school I see a blacked-out SUV sitting across the road. It's a little out of place but probably just some soccer moms new wheels. I made great time home that day, it felt like it was only about 20 minutes. As I walked up the front stairs the hairs on the back of my neck seem to stand on end, I glanced around as I retrieved my key from my pocket, it felt like I was being watched but I couldn't see anything unusual. I think my project today is making me a bit paranoid, I shrug it off and open the front door and quickly shut it behind me locking the deadlock as it closed.

I make my way upstairs and drop my bag in the corner before making my way back downstairs to find a snack to keep me satisfied during tonight's tasks. I go for an old faithful Coke and a couple of bags of snakes. Kind of my go-to for nights like this when I want to have a bit of a sugar buzz to ensure I am on my game. I might give myself an easy task first up and upgrade the jocks fortnight account. I went over to my desk and opened the lid on my laptop but before powering it on I leaned over to the back right corner of my desk, with a bit of a push one of the boards slide out a few cm's revealing a small opening. I reached in and pulled out a small black bag. I looked through the contents before finding my desired flash drive. I take it out and return the bag to the opening in the desk before sliding the board back in place.

I plug the USB drive into the laptop and power it on booting the unit from the Debian instance on the drive that I have customised for particular tasks such as this. As it fires up, my logon scripts start to run, most of them have been written in python to automate the anonymisation of my system. They essentially generate a random mac address, route my traffic through multiple VPN and anonymisation services. It's almost as though I was never here at all, I have got the configuration working so well. I fire up one of my tools and execute the shell connection I have for fortnight, this isn't the first time I was asked to do a bit of a power upgrade to a players account. After a few minutes I had turned the account into an almost unstoppable force, there won't be anyone who will be able to stop him unless they all gang up on him at once. Seems a bit pointless to me to want this but who am I to judge.

Now that the school homework is done it's time to get down to my real work. I power off the machine, disconnect the USB drive and go back to the black bag in the desk. I have created a custom machine for this purpose which is for nothing other than this purpose. Once the task has been completed I will dissolve, erase the drive and rewrite it several times before hitting it with a hammer and then socking the broken up pieces in saltwater. I will normally then take the pieces and drop them along a walk in the city, the pieces are so small that no one sees them but it scatters them around so well that even if someone knew what I was doing they would never be able to piece it all back together. You never can be too cautious. Especially with the type of people, I am about to piss off big time.

I power it up and check over the configuration, I don't need to check it as I have done so on at least three occasions before today to ensure it was ready. I load up my terminal and initiate command and control of my agents. I execute the attack sequence script and pause looking at the blinking cursor on the screen, then after a moment a prompt pops up asking if I wish to execute the sequence. I hesitate. There is no going back from this. I hit enter and watch the sequence execute. All of the machines have been trickling all their data to a cloud platform for weeks and it was time to share the access with the AFP and then destroy everything. Make the house burn down so to speak.

I can only imagine what the people on the other end are going to be thinking, they would be in panic mode right now. Their systems would be locked and my modified version of the eternal blue ransomware will be locking everything down and erasing the data. The Kicker though is the sequence is then designed to max out all abilities of the systems and burn out the components. This will not only bring down the systems but will cost them greatly with all the hosted data centre equipment burning themselves out. I wonder if they will be smart enough to pull the plug before they let everything disintegrate.

Data flow to the cloud services has stopped and I start to see agents going offline. It's nearing the end of the road. I wonder if I will see any chatter about this over the coming days, will it even be talked about. I think it will probably be just swept away like it never happened, they wouldn't want to lose face about allowing something like this to happen. I know you are probably thinking that it won't make a difference, but if I can slow them down and save one girl from being taken advantage of by these scumbags or even slow down the distribution of the poor quality drugs they keep selling then I have saved lives today. That's how I am justifying it to myself anyway.

A message pops up "sequence complete". No point stressing about it now, the jobs done. Clean up time. I execute the sanitisation sequence on the drive and the laptop reboots into a new platform. After about 20 minutes the drive is completely erased. I unplug the USB drive and boot the machine into windows, I open up my homework and then head downstairs to the garage. It's smashing time.

Once the drive has been adequately pulverised I collect the pieces in a small sunglasses bag that I will keep it in until I can take a walk to get rid of the evidence.

49

FORESIGHT: SUSPICION

It's 7 am and that dreaded alarm goes off, I wack it a little harder than I probably should have. I probably stayed up a little late with my celebrations for last night's project. I sat up till around 1 am watching a couple of old faithful movies, Hackers and then Varsity blues. I really should have called it a night after Hackers finished but I was a little spooked after I heard some noises downstairs, I had a look around but couldn't see anything out of place. Anyway no point getting caught up on what I should and shouldn't have done, it's time to crawl out of bed. I need to make myself presentable for the world and grab some breakfast.

Time goes by quickly, I have a shower and breakfast as normal. I throw my daily textbooks in my school bag and put my phone into the front pocket with my keys. I turn and head to the door to head off to school but as I go to open it I freeze. The deadlock is open. I could have sworn I locked that when I came in last night. I stare at it for a few seconds, I mustn't have. I go through the front door and pull it closed behind me.

The morning goes by pretty quickly with not much of note happens until I was sitting in English just before lunch and I look out the window. That same car is parked out just on the other side of our football oval. I can see it very clearly from the second-floor window of the classroom. I see a slight glimmer of light coming from the passenger window like the sun caught a phone screen or camera lens or something. I keep looking at it thinking that something just isn't right, I don't know exactly what it is but I will need to keep an eye out for the SUV in the future. If I can get the number plate I will be able to see who it belongs to. Hacking the transport authority will be child's play but it will set my suspicion at ease.

Suddenly I hear my name "Samantha, what are your thoughts on the required reading?" I was a bit caught aback by the question, I wasn't paying attention to what was happening in the classroom. I sheepishly said, "Sorry miss, I wasn't following". The teacher responded "I know Samantha, that's why I asked you.

Please pay attention to what we are discussing. Have you at least read the book?" I shook my head "not all of it yet miss."

She looked annoyed at me but just shook her head and continued with what she was doing before she chastised me. After a couple of minutes, I glanced back out the window and saw the SUV was gone. Maybe I am just being paranoid. Either way, I will just keep an eye out for it again and see if I can get the plates. I tried to pay better attention for the rest of the class but I still found myself gazing out the window now and then. The topic just doesn't entertain me.

Burrrrt, the bell sounds and I pack my stuff up to go to lunch. I go to the canteen and buy some lunch, then head over to sit with the other girls. Kate is one of my only real friends and she looks up and smiles when I sit next to her. We just go about eating our lunch with minimal chit chat, which is our thing we aren't ones for a lot of talking. It's probably why we get along as well as we do, we both don't mind a bit of silence. The awkward pause in the conversation that drives many crazy is our happy place.

Out of nowhere, the senior from yesterday walks up to us with a sheepish grin on his face. I look over at Kate and she looks at me then looks at Michael, well I think that's his name anyway. "Thanks for yesterday Sam, it helped" still with an awkward grin on his face. I look him in the eyes "That's great, can I help you with something else?" Kate knows a little about my night activities so I don't need to hide it. He shifts his balance from side to side for a couple of seconds and then awkwardly stumbles over his words, "well Umm, I was hoping you might want to go to a movie with me over the weekend?"

At that point his face kind of turns a warm red colour, I think he is a bit nervous. I look over to his right side and I can see some of his friends laughing and whispering something behind him, they must know what he was coming over to ask. The silence stretches for almost a minute before he stumbles over the same sentence again. I look over at Kate and see that she is doing everything she can not burst out in laughter, so she isn't going to be any help to me.

I look up and see that he is getting more and more fidgety, if I don't give him a response soon I think he might die of shame. Strangely, for the first time in forever, I'm not entirely repulsed by the idea. What the heck it's only a movie right. I look up at him "It's Michael. Isn't it?" he nods and I continue "Okay but I get to pick the movie" a grin from ear to ear starts to form on his face and I instantly hope I haven't made a mistake. He reaches into his pocket, pulls out his phone and unlocks it then hands it to me "Can I have your number so I can text you about it?" I take the phone and put in my number and hand it back to him. He stands there for a few seconds looking at me and then realises that it's getting a bit weird before blurting out "okay great, I will text you later. I better go". I just half-smile and nod. I am so going to regret this.

As soon as he starts to walk away, Kate bursts into laughter she is enjoying this, probably a little too much. Once she recovers all she can manage to says is "he's hot if you don't want him I will have him" I just raise my eyebrows at her and ignore her why she keeps rolling around with laughter literally. The rest of the day went by pretty uneventful with Kate giving me a hard time about Michael asking me out, honestly though I don't even know why I said yes, I think I felt sorry for him a little with how awkward he looked. It's only one date and then I can just go back to ignoring him.

The school day finally came to an end. I started to walk home again as normal, responding to a couple of texts from Michael about going to the movies Friday night. I just realised I am going to have to tell my father I am going on a date, that's going to be a fun conversation. I took a deep breath and tried to calm myself, it's surely not going to be that bad. I hope. As I go around the corner in my street I see a glimpse of a black car parked up the street from my house. I take care not to look at it directly. I take a subtle look at the number plate, well at least I think it is subtle and make a mental note of the details. I am going to check who owns it when I get upstairs, maybe my criminal friends from last night are onto me.

I go inside and go straight up to my room and get out one of my Debian machine drives from within my desk. I get straight to work finding my target at the transport authority, I look through the company employees on Facebook and LinkedIn. Suddenly I have a target. Felicity, in administration. I send out an email to one of her accounts with a quickly put together sales promotion for a bookstore, one that she talks about a lot on her socials and offer a 30% deal on the book she was just this morning saying she wanted to read. She will think it's her lucky day until the promo code doesn't work on the site.

She gets the email and clicks on the link, as soon as she does that it loads my command and control system before then redirecting her back to the real book store which doesn't have a sale on the book. I wait for her to look around for a few minutes before I execute the webcam and mic on the machine so I know when it is safe to check out the machine. It's probably thirty minutes before she leaves the room and another thirty before I decide it's safe to poke around. She still ended up buying the book even without my fake discount.

Bingo, she has remote access configured for work with saved credentials. I pull the saved config onto my virtual machine, clone her IP and mac address so that when I connect to the remote location it will look like Felicity. I connect on and open up the registration system, as it loads I finally consider that I probably shouldn't have broken into the transport systems but it's too late now. I am already in. I might as well finish what I started. I punch the rego plate into the systems PKR-601 and wait for the response. After a few seconds, it responds with an unlisted plate, government-owned. I think at that point I turn a slight shade whiter. Am I finally being busted, have I made a mistake. Do they know everything I have done? What am I going to say to John? Oh, this is not good.

I take a deep breath again and just slowly make my way over to the window to see if it is still parked down the street. It was gone though. I back out of the systems and clean my tracks as I go, then power down the laptop and put the flash drive back in the secret spot in the table. I wonder if they were at our house last night. BUG... oh crap, what if they have bugged our house.

I take out a frequency scanner from my draw, I know it's weird that I have one. I do though, so might as well take advantage of it and sweep the house. I checked every room and nothing. Maybe I am just paranoid and they weren't even following me. Maybe they are just a government employee with a company car who just visited someone they know down the street. That's probably more likely than some government cyber squad on the hunt for me. I have always been careful, they wouldn't even know I exist.

I need to get rid of the evidence from last night though, just to be sure. I got it, date night tonight.

Brilliant. I take out my phone and text Michael that I would like to catch a movie tonight. He takes seconds to respond "I would love to, text me your address and I will pick you up". I text him my address and after about 20 minutes he arrives at my house in what looks to be his parent's car. I walk out the door as he gets out and I pull the door closed. I glance around but don't see the SUV. He opens my door to get in the car and I smile at him.

When we get out at the cinemas I retrieve the glasses bag from my pocket and slowly start to drop pieces as we walk down the street. By the time we reach the front the bag is empty and we enter the building. The date went quite well and I even let Michael hold my hand during the movie. I surprised myself a little with that. I am not even sure what movie we saw as we talked softly through most of it, some of the other moviegoers were not impressed about that though we got shushed a few times.

A few days go by without seeing the SUV again. I had been texting Michael a little and had to admit to John that I had gone on a date. He looked quite pale when I told him. I don't think he is ready for me to date yet. He insisted that he meet him but I said maybe once I know it's a thing or not. He grudgingly agreed and things continued on this way for a couple of weeks until I got called to the principal's office one day during classes. I wasn't sure where this was going but I have a bad feeling.

50

FORESIGHT: THE PRINCIPAL'S OFFICE

It's the first time I have been to the front office of the school for anything other than paying for my school fees. I even had to ask at the front desk where the principal's office was. The chairs outside of the office look old and worn. I guess I must be one of the rare students who doesn't have regular visits.

I have been sitting here for about 20 minutes, I can hear someone inside being chastised about their poor judgement and how it will likely ruin their chances of getting into a good college. I guess whatever it is that I am here for it isn't going to be pleasant. I fiddle around in my bag to see if I have anything to snack on while I wait but I don't find anything that interests me, just a blackening banana that has probably been in my bag a little too long. I look around for a bin and see one in the corner, I toss it at the bin and it makes a weird thud noise a little louder than I anticipated. The girl sitting across from me reading a magazine gives me a strange look before continuing to read.

Suddenly the door opens and a boy from my year exits a little too quickly and almost runs straight into me. He looks me in the eyes and I can see he has been crying. The deputy must be a real ball breaker, this should be fun. The book pauses for a few seconds, sort of pulls himself together, straightens up and turns toward the hallway leading out of the waiting room. "Samantha, come in please" a little louder than I was expecting and I jumped a little. Pull yourself together, don't give anything away.

I stand and walk through the door and I see a small framed lady probably in her early forties maybe, which is younger than I expected. She is a little stuffy look-ing but quite pretty really. She gestures for me to sit in one of the two chairs in front of her desk, so I do as directed. I look around the room as I sit down and it a very minimalistic look with very few personal items or personalisation of the room at all. It's almost clinical. I look up after a few seconds and realise that she has been watching me the whole time and it makes me feel a little awkward if I am honest. I just want to know what this is all about so I can take my punishment and get out of here.

She keeps looking at me for another 10 seconds holding my gaze, possibly seeing if I would drop my eyes first but that isn't something I would do even if it makes me feel a little uncomfortable. "Do you know why you are here Samantha?" I look at her for another few seconds before responding "It's just Sam, no one calls me Samantha except my father". She looks at me and I can see her almost deciding what path to take with me and then it looks as though she decides to take an easy tone. She turns her chair slightly and looks out the single window in her office.

She looks as though she is pondering something for a moment, just staring out that window when she turns suddenly back towards me, reaches for something under a pile of books. "I think you should take a look at this" and then turns back towards the window before continuing "I was made aware of this competition coming up next week in Brisbane that I feel you should consider competing in. I glance down at the piece of paper and I can see that it is a government - Hack the World event. I have seen them being discussed on the news a few months ago but I thought they were a bit stupid. Why would I want to go to one of these competitions and put a target on my back by the feds? Doesn't make sense.

I look up and try to put a confused look on my face, "I don't understand, sorry. Why would I want to compete in this competition?" She turned back to face me again and after a few moments all she said was "We both know you have the skills" I was a little thrown by the statement, what does she know. I need to be careful how I respond, I don't want to give anything away. She looks at me and grins. Now I really want to know what she knows, it can't be good. "I have worked here for many years and I have never had you in my office. From what I can see in your school record all I can see is that you are the clear definition of average. I think it's all a lie" she pauses again for a few moments and just as I am preparing to answer she continues "today I was told that you have quite a gift, you are the person that can get things for people. Whatever they need, you can get it. You have a true skill with these computer things but you haven't taken a single computer class. Can you tell me why that is?"

I wasn't sure how I should answer this but I figured short and honest would be best "I haven't taken any units because I wouldn't have learnt anything from them" and then continued to read over the flyer. We both sit in silence for a few minutes considering our positions, well that's what I was doing anyway before I said "have I done something wrong?" she looks up and stares at me, not in a hostile way but more of a curiosity. "I have two choices with you, Sam, I could punish you for the illegal things that we both know you have been doing for students or I could take the advice I was given and let you represent our school in this competition next week. What do you think I should do?"

I looked at her, not sure exactly how I got into this situation but figured that I was in it so I needed to figure out the best option. After a few moments, I decide "I think the competition sounds like the best option but my father will never agree to me going. I don't feel it's something we can afford either" She smiles at me again in a way that almost says I have got you now. "I have already discussed the competition with your father John and he has agreed to you representing the school. He also said that he was quite surprised at the apart abilities as he was never aware that I was so gifted" oh crap, they have already talked to John. I am so dead.

She continues "As for the costs, everything will be covered by the school. It won't cost you or your father anything. However you could win $10K for the school and your choice of new computer system for yourself to the value of $10K also" Now that sounds pretty sweet, I could do with a power upgrade and this way I could buy the best and John wouldn't get suspicious. Sounds like a win, win situation. I get out of this bind I am in, get the school some cash and get a new beefed-up computer. I just have to be careful not to give too much away during the competition or people will know about my extracurricular activities.

"So what have you decided Sam? Do you want to take the red pill or the blue pill?" I almost burst out with laughter with that comment, a teacher who actually has a matrix reference. She continues slightly amused herself by the looks "red pill, I suspend you and tell your father you have been found to be doing illegal favours for students or the blue pill and you go down the rabbit hole by going to the competition?" I couldn't help but giggle a little and I didn't need to think about it was a no brainer really. "I only see one choice really, I will take the blue pill"

She smiles at me again and stands up, "I knew you would make the right choice. I will have everything organised and have the travel details sent to your house so your father can see everything as well" I nod and she reaches out with her hand and I take it. It was a firm shake, she was a smart woman that I could tell already and she is used to getting what she wants. She releases my hand and then gestures to the door. "I look forward to hearing about the cash injection from your win Sam, Do us proud."

I turn and walk out the door still a bit confused as to how that all came about and how exactly I ended up getting signed up for a capture the flag event in Brisbane. Tonight I am certain I will have twenty questions from my dad about my mysterious hidden skill.

I will need to be careful about what I say otherwise I might lose some of that freedom I am used to.

51

FORESIGHT: THE COMPETITION

I am just laying on my bed looking up at the roof letting my mind run crazy with possible scenarios, what obstacles would there be for me to overcome? Would there finally be a system I can't break into? The idea I would not be able to defeat it excites me, I want a challenge I can't overcome, something to make me dig deep and put in everything I have to overthrow the obstacle. I do a sort of internal shrug, who knows what it will be but I need to remember that yes I can have fun with this but I can't reveal myself to anyone. I need to stay safe, John's safety depends on it as much as mine does.

I look over at the alarm clock and it flicks over to 6 am, I am never awake this early. I must be excited about the challenge today. I hear a noise downstairs as John starts to make some breakfast, he told me last night he was going to make me a full English breakfast so I had all the brain food I needed for the competition. He has been pretty excited by the whole idea that I am some kind of computer genius, I don't think I am anything special but I just let him enjoy the moment. I think he deserves that at least for what he does for me. If I win the competition there will be no changing his mind.

I get dressed, pack up my laptop and make my way downstairs to have breakfast. I deliberately left most of my best tools and Linux distros in the compartment in my desk. I don't want to use any of the same tools or scripts I have used in my previous hacker activities. I will do it blind today and have no connection to any of that part of my life. That should make it a bit harder, so I don't draw too much attention to myself. Foresight needs to stay a secret. I walk into the kitchen and my dad looks at me and gestures towards the table, "sit down it is almost ready". I do as he asks and sit my laptop on the end of the kitchen bench, as I do he looks over at it "so that's where all the magic happens hey?" I just look back at him and smile "Something like that Dad".

We sit eating our breakfast in relative silence with some idol chit chat now and then but nothing of note.

We pack up the kitchen, switch on the dishwasher and then we pile in Johns truck to make our way down to Brisbane city. The competition starts at 9 am but John said we should get a head start as traffic is always bad heading into the city, he was right. I think a snail could have moved faster than we could down the highway in some places. When we finally get there, John drives me up to the front entrance before telling me he will find a park and for me to go get myself all sorted inside. He would find a spot to watch so he doesn't cramp my style. I roll my eyes a little at that last point but don't comment, I will never be embarrassed by John.

I get out of the truck and walk through the front doors of the convention centre, wow there are people everywhere. I am a little surprised, this is a big deal. I was expecting about 100 geeks and some family members to support them. All dungeons and dragons playing geeks, you know who I'm talking about. I should know better than to stereotype people though, I am one of those walking contradictions. I am nothing like what most people would think of when someone says, hacker. I look around and yes I see a couple of true geeks but most are just normal looking people.

I start to weave my way through the crowd and after a few minutes I see the registration desk, I turn and head towards it. When I arrive at the desk I can see a line and join the queue to wait my turn. After a few minutes wait, I am next to the desk. "Can I help you miss?" I reach down and pull out the paperwork I was given from the school with my registration and hand it over. She looks at the documentation and then looks back up at me "Do you have any Id?" that's a bit strange but I reach in and pull out my driver's licence, I got it a few months ago but don't use it. My dad said it will come in handy for when I get a job or go to college.

She looks it over and then hands it back before going about her business for a few more minutes. I look around why I wait for her to finish and don't see any familiar faces. However, over on the far back part of the room, I see a couple of government looking people talking to someone that looks like he is military. As I look over at him he catches my glance and we lock eyes. He holds it and doesn't avert his eyes, he is a confident proud soldier who oozes authority. He looks like he is in his mid-50's and has had a hard life. You can just see it in his face. It starts to feel a little weird so I turn and look back at the lady working away at the registration desk.

She gathers up a few things and starts to make her way back over to me. "Samantha, please make your way down to that room on the left where the main competition will start at 10 am" She points to the room past where the soldier was standing, he is gone now. "You will have five hours to complete as much of the competition as you can complete. The rules and instructions will be provided to all participants before the start of events" I nod and take the identification badge and put the lanyard over my head. I collect together the rest of the paperwork and make my way towards the room.

I walk up to the security guys at the door and show them my ID and they let me through to an usher who directs me to my designated desk that is labelled with the school's name. I look around and there are around 30 more desks with 3-4 seats at each one with other schools marked on them. At my desk only one seat is placed so I won't be getting any teammates which I am pleased about, I prefer to work alone, with fewer eyes looking over what I am doing.

I pull out my laptop and start getting everything set up, as I am plugging it into the network I see the military guy from the main area come through the same door I just entered through. He is surrounded by several bodyguards from what I can tell. He sees me and leans over and whispers something to one of them. They turn and look at me before turning back and nodding to him. He turns and walks towards the main stage at the front of the room. The bodyguards follow suit except for the one who looked over at me. He turns and heads towards me, I am not sure what he wants but it freaks me out a little.

When he arrives at my desk he pauses in front of me, he waits till I lift my gaze to look at him. "Excuse me, Samantha, I have been instructed to wish you well for today's competition. The general has indicated that he looks forward to seeing what you can do and that he has heard interesting things about you. He would like to see if it's all true or not".

I am a little surprised at that statement, a general knows about me and what I can do. I don't know if that is a good thing or not. Too late to back out now, John will be very disappointed if I drop out of the competition. So I only have one option "Thanks, I will do my best to impress the general" with a bit of a goofy smile. The human shield nods at me "I will let him know" then turns and walks back towards the general. Wow, he took that very literally, he did not pick up any of the sarcasm that I was laying on thick. I should be more careful about what I say to people, I shrug it off and continue with what I was doing before the interruption.

A few minutes later the room starts to fill with a flood of competitors and their supporters. I see John come in and head towards the back of the room, he looks a bit like a deer in headlights, not sure where to look or go. He sees me watching him and he smiles, straightens his shoulders I guess to make me feel a bit better about the whole situation. Surprisingly even though I know it was fake and that he did it for my sake it does make me relax a little.

Tap Tap Tap, Is this on. Someone was on the main stage and they started to discuss the proceedings of the day. Some general rules of what is acceptable and how long we had to achieve the target milestones. They also talked about a secret level for anyone who finished the competition early, a way to double the prize pack. I have an uncontrollable urge to beat that target, crush my opponents and walk out with the double prize pool. They go on to comment about being grateful for the support of General James O'Connor from the Australian Defence force cyber command for this event. I still think it's strange that General James has taken an interest in me, I hope that doesn't become a problem in the future.

After about ten minutes of fluff, they start the clock "your five hours starts now" I smile, I am looking forward to this. It's time to dig deep and smash my way through these systems, there is no reason to be stealthy and clean my tracks I just need to reach my target. That double prize bounty here we come.

52

FORESIGHT: CRUSHING MY OPPONENTS

I look around and see my competition start a frenzy of activity. They are all teams of three or more and are all banging away on their keyboards. I see people using windows and some on Linux distros like Kali or some other popular pentest distribution. I have even seen a couple of tablets with Nethunter, I don't know why they think that will be useful but each to their own. I realise that I have been sitting here looking around seeing what others are doing for about 15 minutes, maybe I should focus a little and get to work.

I take a deep breath and look over at John, he meets my gaze and smiles while giving me a thumbs up. I think he mouths to me "You got this" he does have faith in me. I take another deep breath and boot up my machine, let's get this show on the road.

I open a terminal on my machine and kick off a network crawl. I want to know what I am getting into here. I see a list of systems like SQL databases, web-servers, PCs and some miscellaneous servers. I can see the versions of software with obvious vulnerabilities. Maybe this isn't going to be much of a challenge after all. I work my way through the systems, knocking them over one by one until they are all done except for one pc. I smash my way through its protections and just look at the screen that was too easy. I look up and glance around the room. Everyone else is still in their little huddles trying to work their way through the obstacles set out for them. As I continue to look around I see the clock at the front of the room and it still has 3 hours and 38 minutes left. That only took me a little over an hour.

It can't be this easy, there must have been some-thing I missed. I take a closer look at the final desktop, why would they make this machine the final obstacle in the challenge? It was simple, it hadn't been patched ever and was just a basic Windows 10 system. What am I missing...

I check through the network configuration and see that it has a dual IP, maybe it is a gateway to an-other isolated network. I load some tools and do a scan of the network range, there are some servers just sitting there. I check them out cautiously, something about this isn't right. I load up Wireshark and monitor the traffic in and out of the server's bingo. They are honey pots ready to distract me from the real surprise 192.168.10.189 the IP address of where all the data is being sent back through too. I check it out and there is a firewall between me and the end device. I do some light touch work on the firewall and it turns out that it is cisco, one that hasn't been patched.

That's my in, I check to see if it is vulnerable to any known attacks. I attempt to access the web portal and I see that it is vulnerable to credential caching. If I can execute an export command it will allow me to pull the last admin login details used. I put together a rough script to do what I need and pause. I don't think this is the extra part of the game, I think that this is the back-end systems that are being used to monitor the game. Do I stop or do I go for it, crush my opponents and the people who are running the competition. I consider it for a moment, screw it I want that double prize pool.

I execute the script and pull the admin credentials. I log onto the firewall and reveal the VPN credentials, I use them to gain an active connection to the network and do a soft crawl and this is defence gear. Oh crap, I don't know if that was a good idea after all but this is the name of the game isn't it, crush all obstacles in my path. Capture all flags and systems on my way. I make my way through the network and find my way in with a weak set of credentials. I logon to the user and decide to have a little fun.

I load a screensaver on the main pc that is displaying the event details on the main stage, it's there to display updates for the event. I load a video that will have a skater chick who lands an awesome trick and then rushes the screen shouting you have been owned. After a few seconds of pause, I decided to add a tag line with my school's name on it at the end and execute. I pull back from the systems and just relax in my chair with a bit of a goofy grin on my face.

I look around and see the general looking over in my direction and he is talking to one of his team. He nods at me with a sort of acknowledgement. They know what I have gotten access to, maybe they are trying to figure out what I did. Suddenly the video plays on the big screen and a smile creeps onto the generals face. I look over at the clock 2 hours 58 minutes remaining. I see him signal for one of his team and suddenly the video disappears. My school suddenly jumps to the top of the leader board with a time of 2 hours 2 minutes listed next to it. With a bonus round star tagged and a blinking winner icon.

A few moments later someone comes on stage and informs everyone that the winner of the competition has been decided and places two and three are still up for grabs. Keep working hard on the objectives. I look over to John and he has a blend of astonishment and pride written all over his face. I think there will be no changing his mind that I am some sort of computer genius now. I smile at him and he nods at me mouthing "nice work, I am so proud of you" it gives me a bit of pride at that moment and I forget about the unwanted attention that I have probably just brought on myself.

I glance toward the general again and I see him whispering something to the human shield that talked to me initially. He started to make his way towards me, here we go. When he arrives in front of my desk he pauses again waiting for me to look up at him. "Samantha, the general would like to have a word with you in private for a moment. Please follow me" the look on his face didn't seem to me as though the request was voluntary. So I rise and follow him to a door at the side of the main room. He opens the door and as I walk through the general is standing to one side talking to the guy that he asked to pull my video from the main screen.

He meets my gaze as the door is pulled closed behind me with a bit of a thud. "Miss Erkhart, quite a show you put on today" I just look at him for a few moments and nod. "I had heard you were good but 2 hours 2 minutes and you breached the private network that wasn't even supposed to be part of the game" he walked over to me until he was less than a few feet away "what gave the honey pot away, why did you change your target? We could see you checking it out then just like that you changed direction". I looked around the room while considering my answer everyone in the room was waiting to see what I would say. "Your concealment of the logging traffic was bad, you should have only been transmitting major alerts back to control and stored the rest on the device for later review. Too much traffic was unusual, it just felt off." I looked back at the General, "I think your team could have made the objectives a bit harder don't you think?" I heard a slight chuckle from someone behind me at that point, who received a sharp glare from the general. It stopped instantly.

"Miss Erkhart, let me tell you something, the 29 other teams out there haven't even passed stage 4 and you have taken out the entire 10 stages, pushed into the bonus network as well as broke through to the secure network. I think the only person who found it easy was you." He reached for something inside his coat jacket and pulled out a card. He handed it to me. I look over it and all it has is a phone number. "If you would like to discuss an opportunity for after you graduate in a few weeks give this number a call, tell them the general sent you" he looks over at his man behind me and nods. The human shield opens the door behind me and tells me it's time to head back to the competition.

As I walk through the door I see John arguing with one of the bodyguards asking why they are talking to me in the room, he looks up and sees me. I smile and wave at him. He stops and comes to me immediately "are you, okay honey, what was going on in there?" I look him in the eyes "They were just interested in my techniques and how I defeated their defences so fast, I just told them it was because they didn't really have any defences, maybe they should do better next time" that set my dad at ease, his shoulders seem to relax a little. "Do you want to wait with me at my desk until everyone else finishes?" he nodded "Of course I would, it will put me closer to the action and will make it easier to take some pictures of you getting your prize".

It has been an interesting day, my head was spinning with what has occurred and the card the general had given me was almost burning my skin. I wanted to call and find out what was on offer but I won't be calling, it's too much of a risk. Or should I, oh I don't know.

53

FORESIGHT: TAKING HOME THE LOOT

I started to pack up my computer equipment while I waited for the competition to end if I didn't have to wait for the prizes I would bail already but I don't have much choice. If I want the $10k in cash and $10k voucher for new computer equipment I need to stay until they hand out the prizes. I think the Principal should be happy I have won the school $20K worth of school equipment. That should help to bring the computer lab into the 21st century. Maybe they might name the lab after me, that's an odd thought. I truly hope they don't do that.

I must have been just staring off into space, as when I look over to John he asked if I was okay. I nod "Yes

I am fine Dad, I am just daydreaming, thinking about what kind of new computer system I can get with $10K" he smiles. "Dad, I am glad you came with me today, it made me feel at ease and want to do my best" John looks at me with the biggest smile on his face "It was my pleasure, Sam, I would never have missed this" we continue to look at each other for a few moments longer than I continue packing up. As I put the last few items in my bag, I notice I am being watched. As I look around the room quite a few of the other school teams are just outright staring at me, it's getting a little weird.

Another hour passes, the competition finally comes to an end and the competition winners are announced officially. We are brought up on stage and given our prize packs while constant flashing of the camera's record every moment. I smile and go through the motions as required but as soon as it is socially acceptable I return to John "Dad, let's get out of here!" he nods and we gather everything up and start to head for the door. The main foyer area that was filled with people this morning is now more like a ghost town than a bustling event, I guess everyone got what they came for. John points towards the elevator and we head down to level four of the car park.

As I exit the elevator I see something in the far corner of my vision, it's that black SUV that I have been seeing around the school and in my street. I don't want to look over in case I give myself away. John points towards the back corner of the car park and tells me the car is parked over there, I barely acknowledge his statement as I am too concerned about that SUV and what is it doing here? Am I still being watched?

I pick up my pace a little to get to the car as quickly as I can without making it too obvious, John notices though and looks over at me a little puzzled "are you in a hurry to get somewhere?". I look back at him and consider if I should tell him about the SUV or will that just freak him out. I will keep it to myself for now but if it gets out of hand I will bring him in the loop. "There is a great computer store on the way home and I was hoping we might get there early enough to check out some upgrade options", he instantly smiled "of course. Anything for my super hacker genius daughter". I just smile at him and pick up the pace a little more and he does the same.

We reach the car and we both pile in. I grab my seat belt and fasten myself in. I look back over to the SUV as we start to make our way out of the underground car park. I can see movement, there are several figures in it but the tint is too dark to make anything particular out. I hope it's nothing to do with what I did the other day, if I have put John at risk I will never forgive myself.

As we start our ascent out of the car park, the lights on the SUV switch on. Someone has just started the engine and I see in my side mirror that it has started to pull out of the park. As we continue to make our way out I see several flashes of light from the SUV behind us slowly making its way behind us. We finally reach the final exit gate and my father put his window down to insert our parking card, John prepaid for the days parking that morning. As he does his thing with the gate I see the SUV pull up to the gate beside us, the window starts to move down, this is my chance to see who it is that seems to be following me. I can see the top of the drivers head appear with a brown colour to it. Here we go. Suddenly John lurches forward and the view of the SUV has gone. No, no, no, that might have been my only chance to find out who it is.

Argh, I guess it wasn't meant to be. We pull out of the driveway and John starts to make his way towards home. He turns to me after a few minutes "Let me know where you want to go hacker shopping?" I really won't live this one down now, maybe the new computer wasn't worth it, after all, I think I preferred it when John had no idea about my skills. I will probably be stuck fixing all of his work friend's computers now as he will more than likely brag about how great I am with computers.

I just nod in response to his statement and we continue to drive in relative silence for almost 30 minutes before I give him some direction to the computer store. When we pull up in the car park I glance around to check for the SUV, please, please make it that I am just paranoid and it isn't following me. I take a deep breath and almost sigh with relief when it is nowhere to be seen. I am just delusional and maybe a bit paranoid. Maybe I just need to get some more sleep at nights instead of doing dangerous things as Foresight.

I walk around the store almost like a little kid in a candy shop, I can get some much with my prize money. I put a bit of a list together in my head before looking over to the guy at the counter "can you help me place an order?" he gives me a bit of a look over and says "what sort of thing do you want to do honey?" Honey, oh that's not cool, he continues "play music? Social media? Maybe watch some movies or YouTube or something?" Oh, this guy is going to cop it. I look over at John for a moment and he has this look of anticipation like he is waiting for the bomb to explode. He knows me way too well.

I look back at the guy behind the counter and see his name is Jack. "Jack is it?" he nods "Do me a favour would you, pull your head out of your arse for a moment and listen." The look on his face suddenly changes, he doesn't think he is so smooth now. "Just because I am a girl it doesn't mean that I don't know anything about computers, I just won the prize that this stores head office donated to a capture the flag contest in the city. If you call me HONEY one more time I will make sure that everyone in this state knows what you get up to on your computer when you are alone." I feel a little bad right now as he looks as though he is either going to cry or worse wet himself. I look at him for a moment longer before continuing. "Now do you need to write my order down?" he nods and grabs a pad and paper.

"I want Asus Maximus XI hero motherboard, NVidia GeForce RTX 2080 Ti video card and I want the best RAM you can get me and as much as you can fit on the board". I add a large SSD and about 10TB worth of additional storage and an Intel Core i9 processor to drive it all. I add a see-through gamer case that can hold all of my new gear and even splurge out on a 49" super ultra-wide screen. This thing is going to be a nice machine.

I turn to look at John and he has a glazed over kind of look on his face, I think I have lost him. I look back at Jack the jerk behind the counter and he looks almost impressed. He finishes writing my list out "we have all of this in stock do you want me to get it all ready for you now?" The look on my face must have answered his question, as he just nods and busies himself with gathering my order. I watch him zig zag through the storage area behind the counter collecting my items and after about 10 minutes he returns with a trolley full of stuff.

He walks over to the counter "Miss, I have upgraded your power supply, cooling system and will add in an LED package for your case for no extra charge. What colour would you like?" I point to the blue kit and he adds it to my collection. "Do you have your voucher number from your prize?" I reach into my pocket and collect the voucher and hand it over to him. "Wow that's an impressive prize, I assume we will be seeing you again, Samantha, you are going to have quite a lot left to spend after this order" I don't think I will need anything else, maybe I could buy John a laptop and they sell big screen TV's that could be a good choice. I think the 32" we have now could do with an upgrade.

Jack puts it all through the register and then offers to help take it to the car which we agree to. We load it all in and start our drive back home. It was a good day today, a really good day especially for John and me. This whole event seems to have been a great bonding moment for us. It puts a bit of a smile on my face as I think about it.

54

FORESIGHT: MEETING JAMES

It's been a few days since the competition now and my father is still hyped up about the whole experience. I could do anything I wanted and he would still be happy with me. He even agreed to me going out with Michael tonight, he said that I had earned a bit of fun will all of my hard work at the competition. I don't have the heart to tell him that to me it was easy. I still don't know how my gift works but it's like I just close my eyes and visualise and the machines do what I want. Like I can control them almost with my mind.

I know it sounds weird but I have been hiding this gift of mine for years and now it's out in the open. Well a little, people just think I am a super hacker or something which is partly true. I am a very skilled hacker but I am not your normal hacker. I am Foresight.

I hear a knock at the door, it must be Michael, he is taking me to see a movie, I don't even know what it is we are supposed to be watching but It will be good to just chill out for a while. I check my hair and outfit in the mirror before turning for the door. I make my way downstairs and see Michael sitting at the kitchen table with John. I hope John is being nice "Dad, I hope you aren't giving Michael a hard time or anything?" a small grin creeps onto his face, I instantly have a churning feeling in my stomach and I can see from Michaels face that he was stirring him up a little. "I just told Michael that I have a shovel in the garage and if I hear that he so much as touches a single hair on your head I will be burying him in my back yard with it. That's all Sam." I fight back a giggle but can't hide the smile which John sees. It makes his smile stretch right across his face.

"Goodbye Dad, I will be back before midnight. Michael lets go before he tells you about his guns" I look over at my dad and his smile gets even bigger if it's even possible, I shouldn't encourage him though. He stops though after a few seconds, "Your curfew is 11 pm Sam, not 12. Make sure your home by 11 please" it was worth a try but I just nod. I look over at Michael and instantly he gets up and we walk out the front door. We look over at each other while we walk down the front path, he opens my door as we approach his car. When we arrive at the cinema he opens it for me again, he is pulling out the gentlemanly acts tonight. Michael might be a good guy, I might have found a good one. I am still not sure I am ready for a relationship though but I am going with it for now.

We start to walk to the cinemas and he reaches to take my hand, I let him. It's strange, I have this weird butterfly sensation in my stomach and it sort of tingles around the areas where our skin touches. I don't understand this whole teenage hormones stuff, this is all alien to me. I would be less scared to hack the DoD than I would be to kiss Michael tonight, I know it is a possibility he will try at some point and I think I am okay with it but it terrifies me a little.

The movie goes by fast, it's some sort of teenage rom-com and I don't even remember the name. I was freaking out the whole time as a few minutes in Michael, moved on from holding my hand and placed his hand on the inside of my thy, high up on my leg. I was experiencing that warm tingly sensation right up the inner part of my leg and it was almost burning hot where his hand was resting. It was exhilarating and terrifying at the same time. I was almost frozen, was I ready for this and was I ready to take things to the next level. Did I want to get more physical with Michael?

The movie finished and we walked in silence to the car, a sort of anticipation that something was going to happen in the air. When we get to his car I see him look at the door and in anticipation that he will open the door for me again, I turn to the side and pause slightly. As I do I think Michael took it as me giving him a queue, a green light? He turns toward me and places his hands on my hips. He gently guides me back onto his car, I feel myself bump into his car coming to rest on it, when I do he slides his hands slowly up from my waist, sliding them under the bottom edge of my shirt. As his hands slide onto my bare skin I can feel the tingling sensation spread across most of my body, I don't want him to stop there, I want him to touch me everywhere. These clothes between us are frustrating me, I want to feel his skin on mine and I want to feel this sensation spread across every nerve in my body. I want to lose control, just a little.

He presses his body to mine and leans forward. He is going to kiss me. I look at his lips as they near mine, they look gentle but determined. They meet with mine after what feels like minutes. They feel soft to touch, moist and inviting. I feel a strong desire to claw at his clothes as he pulls back and slightly parts his lips then locks them with mine again. I feel a little short of breath, I didn't know this is what it would feel like. It feels like a drug is entering my system, like an addiction I know will be hard to kick.

Suddenly, something grabs me from behind, a bag is thrown over my head and my hands are tied behind my back. I believe they are doing the same to Michael, I don't know what is happening. One of my past activities has finally caught up to me. They pick me up and push me into the back of a car, once I am in they secure my feet. I feel another body pressing against mine, I presume that is Michael. They are thrashing around trying to break free but it's no use. We are secured and gaged, we are not going to go anywhere. This is not how I saw my night going a few minutes ago.

I can feel the car moving, I try to pay attention to what I can hear and the turns we take but it's hard to keep track. We have been driving for a few minutes when suddenly the vehicle stops. I hear two people get out and open the car door next to me. They grab my arm and pull me out of the vehicle, they leave Michael where he is and close the door behind me. They cut the zip tie around my ankles and lead me on a short walk which feels like some sort of gravel road for about 50 metres.

I feel the zip ties on my hands get cut as we stop. What is going to happen to me, is this going to be the end. I am free to move though, I still have a chance. Suddenly the bag over my head is pulled off. I struggle to see anything as my eyes adjust but after a few moments, I see the general standing in front of me. It's General James from the competition. Oh crap, what is going on here? I guess they were not too happy I broke their network in front of everyone.

He looks at me and takes a few steps closer "Sam, I have been waiting for a call, it's been almost a week. I am not a patient man you know". I look around and see many of the same faces that were at the competition in his team. I am not sure how I should respond but decide to just go with my usual tact or lack of "I can see that, you could have just asked me to talk, you didn't need to kidnap me you know".

He nodded. "You are probably right but the boys said you were getting a little frisky with your boy over there and they didn't want to watch where that was heading". Oh wow, they have been watching me, how embarrassing. Although it's dark I can feel my cheeks get a little warm, I am blushing. Pull it together Sam. "I think it is time we had a proper talk don't you?" he takes another step forward so he is right in front of me "I think you have some explaining to do, don't you?" he pauses for a moment and just looks at me waiting for me to respond but I don't say anything. "They will take you home now and Joe here will pick you up at 10 am. So we can have that chat, what do you say?" I consider it for a moment "Okay but this time I can put myself in the car, no more snatch and grab deals". The general nods in agreement with a slight smile creeping on to his face "Done. I will see you tomorrow Sam and you should consider toning down the public displays of affection, don't you think?" wow he had to go there didn't he, I am almost dumbfounded and don't know what to say. Joe grabs my elbow and leads me back to the car, still a little aggressive like if you ask me "Hey, watch it, bozo. No need to be so handsy is there?" he just gave me a sharp look and squeezed a little harder. I guess he does.

We get halfway back to the car and someone pulls the bag back over my head, was that necessary. I roll my eyes even though I know that no one can see it. They open the door as we approach the car and shove me back in next to Michael. "Are you okay Michael?" he murmurs something through his gag and then someone gives him a whack in the ribs just hard enough to stop him from saying anything further. The dive back seems to be faster than the initial one, maybe due to me knowing what is happening. I don't know if Michael will think the same though and I don't think I will be able to tell him what happened either, I will need to make something up to keep him out of this.

As the car comes to a halt, someone opens the door and pulls both of us out, throwing Michael on the ground near his car and they lead me over to him. They hand me a pair of wire cutters "Don't remove your masks until you hear us leave, understand?" I quickly respond "yes I understand". I hear them walk away and the car leaves. I wait a few more moments before reaching up and removing my headcover. I look down and immediately move to untie Michael. I cut his legs free first and then his arms. He reaches up and pulls off the cloth bag over his head and gag in his mouth. "What in hell was that, who were those people? What have you gotten me involved in?"

I consider my answer for a few moments before responding "They are a crime gang that wasn't happy with a job I did, apparently I cost them a lot of money. They don't want me to do any more jobs for their competition and only do jobs for them" I hope my story is enough to satisfy his need to know what just happened. He looks angry and scared but it would seem he accepts the answer. He stands up and walks around the driver's side of his car, unlocks the car and gets in. As he does "Let's get out of here", I can't argue with that sentiment and get in the passenger side.

The drive home is very quiet, neither of us says anything but I can see he is thinking hard about something, I guess I will find out soon enough. We pull up outside of my house, he turns off the car. I look over at him and wait. I think we both know we need to talk about what just happened. "Sam, look I like you" I go to respond but he cuts me off "I do but this type of life isn't for me. I can't live my life thinking that someone could just grab me out of nowhere without warning. That isn't me. We had fun and I don't want to be enemies but I think we can't date anymore" he pauses, I can see this is hard for him so I just wait for him to continue "I wish things were different, I do".

I am not surprised I knew this was a possibility, I can accept this. I look at him for a moment and then lean over putting my hand behind his neck and pulled him toward me. He doesn't resist and we kiss passionately for a moment. I pull back and look him in the eyes, I can see he is torn. "Good-Bye Michael". I turn and open my door. I get out, close the door and walk straight back to the house without ever looking back. I reach down to grab the handle when I feel a hand take mine. He pulls me back around towards him and pulls me back into his embrace, we kiss and instantly I feel the warm tingly sensation spread across my body like fire spreads across fuel. After a few minutes of lost time, the porch light comes on.

We both jolt back to reality and Michael looks at me for a few moments "Goodbye Sam". Is that it, are we still finished. Oh, I am so confused.

I turn and reach for the door handle again, take a deep breath and push the door open. I can see John standing with his arms crossed to my right as I enter, "You are late Sam?" I turn to face him and just nod "I know Dad, I am sorry" he looks puzzled, maybe he thought I would argue with him. "If it's okay I just want to go to bed?" he must still be surprised by my answer "Okay, we can talk about this tomorrow night". I nod and then turn for the stairs. What a night.

55

FORESIGHT: CHAUFFER

I wake to the sound of rain on our roof, it's a relaxing and sleepy kind of day but my brain has been running wild most of the night. On one side all I can think about is the kiss with Michael on my porch before my dad ruined the moment by switching on the light. The feeling that pulsated through my body, I can almost feel it still moving through my nerve endings. It's a feeling I didn't know was possible but if I am honest with myself completely I loved it. It was like a drug pulsing through my veins of pure adrenalin something I want again, that I know for sure. I don't know if last night was the end for Michael and me, I hope it isn't.

Or the fact that Michael and I essentially got snatched from outside the cinema last night by government goons. I don't even know how I should feel about that, it is the reason why Michael doesn't want to continue whatever it is that we have going but I enjoyed the adrenalin rush that I got from not knowing what is happening, thinking that it could be over with similar electricity pulsing through my body as to when I was merging Michaels body with my own in my mind. I am not sure which one excited me more, which I enjoyed the most. I should be concerned about the meeting with the general today but I am abuzz with the excitement of what could happen. The danger and possibility are exhilarating.

I push both of them from my mind. I need to concentrate, I need to get out of bed and get ready for today. I push myself out of bed and head for the shower. I soak up the warmth as the water runs over my back, I stay this way for a few minutes before I feel more relaxed and ready to face the day with a clear mind. I get out of the shower and head to my cupboard to pick out some clothes for the day, I better make sure that I wear something appropriate this time, I don't think last night's skirt was the best option to be snatched and thrown into the back of a government car or whoever else decides to snatch me today. Jeans and a shirt would be fine I think, maybe even a nice jacket to keep off the rain. Done and done.

I turn and check the outfit in the mirror, I look good, happy with that choice. I turn towards the door and start to make my way downstairs when I hear John talking on the phone with someone which sounds important but as soon as he sees me he tells the person on the other end that he will call them back in a moment. As I approach him he looks at me and takes what looks to be a deep breath. Oh crap, this is going to be a bad conversation. "Sam, I think we need to have a talk" he signals for me to go sit at the dining room table which I do. He follows suit and then pauses for a moment. I can see he is thinking hard about something, I just wait, whatever it is I will just take the punishment.

He looks up at me and holds my eyes "Sam, I think we need to talk about last night." I figured that's what it was going to be about, I was late and I know it. I have never Brocken curfew before and I assume he will want to be tough on me to make sure I don't do it again. "Okay, Dad" is all I say. He looks like he is going a bit red now, he must be really mad at me. "I think we should talk about how babies are made, I know it's a bit awkward talking to your silly old dad about these things but I need to you understand the risks about what you might be thinking about doing" I start to go as red as a tomato, wow, oh crap. "Dad, we don't need to talk about the birds and the bees. I have had sex education at school I know how it all works. Please stop, please stop talking" I can feel myself starting to go even redder if that is possible.

"Sam, I know it's not something you want to talk to me about but I saw how you and Michael were on the porch last night and I want you to be safe. I want you to know that even if it's awkward I am here for you to talk boys or whatever you need Okay?" I look at him "Dad, Okay I know I can talk to you about it, if I need to I will but please leave it for now before I die of embarrassment" he shuffles in his seat "do you think I want to talk about it either, Okay I will leave you be for now. I am heading to the office, I will probably be late but please don't have Michael over why I am not here" I just nod still a little thrown back by the topic of conversation. Wow, I was not ready for that.

The next hour goes past very quickly and the conversation with John made me forget all about the other stuff. It's 10 am, they will be picking me up soon for our chat. I grab my keys from the bowl and I walk out the front door, as I pull it closed behind me I see a reflection in the glass beside the door. It's that black SUV I have been seeing everywhere, it starts to slow down as it approaches the front of my house and then comes to a complete stop. I take a deep breath and turn to face the car. The front window starts to go down and I see Joe looking out at me. It's been the Generals guys who were following me before the contest. They have been watching me for a while, oh crap maybe this meeting is not such a good idea.

I don't know if I have much of a choice now, I will just go with it and hope for the best. I start to walk down the path and as I get close to the SUV, Joe gets out and opens the back door "Good morning Sam" I nod at him and as I look in the back seat I see someone else sitting on the other side. It's a lady dressed like the other goons. She looks over at me as I get in the car "you will need to put this over your head please, I believe you have worn one of these before?" I look at her and nod "is this really necessary?" she turns back to me "if you don't put it on voluntarily I can ask the guys to make you" wow that's a bit aggressive, I just do as I am asked and as Joe closes the door next to me I put it over my head.

We drive for about 25 minutes I believe before we come to a complete stop, I can hear some strange noises outside the car like we are in a wind tunnel or something. I feel the car move down we must be in some sort of elevator. Once we come to rest I feel the car start to drive forward again. Looks like we must be where we are going. The car comes to a stop and I hear the guys in the front get out. One of them opens my door "you can take that off your head now and get out".

56

FORESIGHT: GETTING THROUGH SECURITY

My eyes take a few moments to adjust after the headcover was removed, I look towards the door and slowly slide out of the SUV. As I do I can see we are in some sort of underground car park, it has fifteen or so black SUVs just like the one that picked me up. I continue to look around and all I can see is an elevator with a security pad on the left of the doors which I will assume is so only authorised people can enter the elevator. Joe gestures for me to make my way to the elevator and I do slowly but as I do I notice something strange. There doesn't seem to be a way into this room I am in. How did we drive down here, does one of these walls move out of the way? There doesn't seem to be a way in or out of this room except for the elevator doors in front of me and the SUV isn't fitting in there.

We walk up to the elevator door and the girl who was sitting next to me in the back seat walks up to the security pad and places her hand on the panel. I see it gets scanned by a bright blue light, it will be checking the palm print and also vein patterns. After a few moments, the panel blinks green and she leans forward and a retinal scanner scans her right eye. After a few more moments the elevator door opens, she gestures me in. I step forward and make my way to the back edge of the elevator, it looks like any other building elevator I have been in except that instead of that it doesn't list the building floors. The girl enters and swipes what looks to be a security fob on a panel and it lights up with 10 level options that all say sub and a number randomly spread between sub 1 and sub 22. There must be at least 22 floors. She selects sub 4 on the panel and the doors close.

The two guys who were in the SUV don't come with us "Joe isn't coming with us?" I don't expect a response but I wanted to break the awkward silence, unusual for me to be concerned about silence. I feel the elevator start to move downwards, oh we must be going underground. She turns to me "Joe is not authorised to enter the elevator" I just nod. It takes only a few moments to arrive at our floor. AS the doors open, I see a long hallway with several cameras on the ceiling. I am gestured to exit the elevator "This is where I leave your Sam, please make your way to the other end of the room and place your palm on the security pad at the other end" I step forward and out into the long skinny room. I stop and look for a moment and then turn as I hear the elevator doors close behind me.

The walls are covered in a mix of different art and I can see looking around that there is a lot of sensors and camera's keeping watch on the room, probably the whole place most likely. Where the hell am I, I hope I haven't made a mistake coming here, I hope this is not my final cell and I never leave. I shake the thought from my mind and take a deep breath. Pull yourself together they are watching. I look at the other end of the room and see the security panel as was instructed, I make my way to it slowly, looking at the artwork as I go. Once I reach it I pause for a moment before placing my hand on the panel as I had seen the girl do at the elevator. The same blue light scanned my hand and after a few moments, a computer-generated voice requested me to place my eye in front of the retinal scanner.

I stood there for a few moments while it scanned my eye and suddenly it dawned on me, how do they have my palm print or my retinal scan information. Suddenly the voice sounded back, temporary access authorised – Samantha Erkhart. Suddenly the door opens in front of me and I see two guards standing holding what looks like machine guns at the back of the room, I slowly enter and I can see that there are at least four more suitably armed guards around the room. This looks like some sort of security checkpoint and as I step forward someone gestures for me to come towards them. "Samantha, please place all of your personal items in the tray and remove all remaining metal objects." I do as directed and I am gestured to walk into a machine in front of me. It's a full-body scanner, these people are not taking any chances. I am in the machine for what feels like five or six minutes before I am gestured to continue through. "You will get your items back when you leave" I nod as I don't see any point in arguing, it wouldn't get me anywhere. They gesture for me to go through the door in front of me, opening it for me as I approach. As I enter the next room a pretty blonde girl who looks to be not much older than me approaches "We have been expecting you, Samantha, please follow me" I follow her down a maze of passages, this place is huge. We finally get to a room that looks like a teenager's retreat, a big-screen TV, a wall full of movies and a selection of gaming consoles. "Samantha, please make yourself comfortable, the general is currently dealing with a situation and will be with you as soon as he has resolved it".

She turns and closes the door behind her, I look around and see multiple cameras watching the room, I walk over to the door and try to open it. I am locked in. I guess I might as well make myself comfortable.

Too be continued...